INTRODUCTION TO ENVIRONMENTAL ANALYSIS

Analytical Techniques in the Sciences (AnTS)

Series Editor: David J. Ando, Consultant, Dartford, Kent, UK

A series of open learning/distance learning books which covers all of the major analytical techniques and their application in the most important areas of physical, life and materials sciences.

Titles Available in the Series

Analytical Instrumentation: Performance Characteristics and Quality
Graham Currell, University of the West of England, Bristol, UK

Fundamentals of Electroanalytical Chemistry
Paul M. S. Monk, Manchester Metropolitan University, Manchester, UK

Introduction to Environmental Analysis
Roger N. Reeve, University of Sunderland, UK

Forthcoming Titles

Polymer Analysis
Barbara H. Stuart, University of Technology, Sydney, Australia

Chemical Sensors and Biosensors
Brain R. Eggins, University of Ulster at Jordanstown, Northern Ireland, UK

Analysis of Controlled Substances
Michael D. Cole, Anglia Polytechnic University, Cambridge, UK

INTRODUCTION TO ENVIRONMENTAL ANALYSIS

Roger N. Reeve
University of Sunderland, UK

JOHN WILEY & SONS, LTD

Published in 2002 by John Wiley & Sons, Ltd
Baffins Lane, Chichester,
West Sussex, PO19 1UD, England

National 01243 779777
International (+44) 1243 779777
e-mail (for orders and customer service enquiries): cs-books@wiley.co.uk
Visit our Home Page on http://www.wiley.co.uk or http://www.wiley.com

Other Wiley Editorial Offices

John Wiley & Sons, Inc., 605 Third Avenue,
New York, NY 10158-0012, USA

Wiley-VCH Verlag GmbH,
Pappelallee 3, D-69469 Weinheim, Germany

John Wiley & Sons Australia, Ltd
33 Park Road, Milton, Queensland 4064, Australia

John Wiley & Sons (Asia) Pte Ltd, 2 Clementi Loop #02-01,
Jin Xing Distripark, Singapore 129809

John Wiley & Sons (Canada) Ltd, 22 Worcester Road,
Rexdale, Ontario M9W 1L1, Canada

Library of Congress Cataloging-in-Publication Data

Reeve, Roger N.
Introduction to environmental analysis/Roger N. Reeve.
p. cm. – (Analytical techniques in the sciences)
Includes bibliographical references and index.
ISBN 0-471-49294-9 (cloth: alk. paper) – ISBN 0-471-49295-7 (pbk.:alk. paper)
1. Pollutants – Analysis. 2. Environmental chemistry. 3. Chemistry, Analytic. I. Title.
II. Series.

TD193.R44342001
628.5–dc21 2001026255

British Library Cataloguing in Publication Data

A catalogue record for this book is available from the British Library

ISBN 0-471-49294-9 (Cloth)
ISBN 0-471-49295-7 (Paper)

Typeset in 10/12pt Times by Laser Words, (India) Ltd.
Printed and bound in Great Britain by Antony Rowe, Chippenham, Wiltshire.
This book is printed on acid-free paper responsibly manufactured from sustainable forestry in which at least two trees are planted for each one used for paper production.

To Rose

– my wife, companion and friend

Contents

Series Preface

There has been a rapid expansion in the provision of further education in recent years, which has brought with it the need to provide more flexible methods of teaching in order to satisfy the requirements of an increasingly more diverse type of student. In this respect, the *open learning* approach has proved to be a valuable and effective teaching method, in particular for those students who for a variety of reasons cannot pursue full-time traditional courses. As a result, John Wiley & Sons Ltd first published the Analytical Chemistry by Open Learning (ACOL) series of textbooks in the late 1980s. This series, which covers all of the major analytical techniques, rapidly established itself as a valuable teaching resource, providing a convenient and flexible means of studying for those people who, on account of their individual circumstances, were not able to take advantage of more conventional methods of education in this particular subject area.

Following upon the success of the ACOL series, which by its very name is predominately concerned with Analytical *Chemistry*, the *Analytical Techniques in the Sciences* (AnTS) series of open learning texts has now been introduced with the aim of providing a broader coverage of the many areas of science in which analytical techniques and methods are now increasingly applied. With this in mind, the AnTS series of texts seeks to provide a range of books which will cover not only the actual techniques themselves, but *also* those scientific disciplines which have a necessary requirement for analytical characterization methods.

Analytical instrumentation continues to increase in sophistication, and as a consequence, the range of materials that can now be almost routinely analysed has increased accordingly. Books in this series which are concerned with the *techniques* themselves will reflect such advances in analytical instrumentation, while at the same time providing full and detailed discussions of the fundamental concepts and theories of the particular analytical method being considered. Such books will cover a variety of techniques, including general instrumental analysis, spectroscopy, chromatography, electrophoresis, tandem techniques, electroanalytical methods, X-ray analysis and other significant topics. In addition, books in the series will include the *application* of analytical techniques in areas such as environmental science, the life sciences, clinical analysis, food science, forensic analysis, pharmaceutical science, conservation and archaeology, polymer science and general solid-state materials science.

Written by experts in their own particular fields, the books are presented in an easy-to-read, user-friendly style, with each chapter including both learning objectives and summaries of the subject matter being covered. The progress of the reader can be assessed by the use of frequent self-assessment questions (SAQs) and discussion questions (DQs), along with their corresponding reinforcing or remedial responses, which appear regularly throughout the texts. The books are thus eminently suitable both for self-study applications and for forming the basis of industrial company in-house training schemes. Each text also contains a large amount of supplementary material, including bibliographies, lists of acronyms and abbreviations, and tables of SI Units and important physical constants, plus where appropriate, glossaries and references to original literature sources.

It is therefore hoped that this present series of text books will prove to be a useful and valuable source of teaching material, both for individual students and for teachers of science courses.

Dave Ando
Dartford, UK

Preface

Interest in the environment continues to expand and develop. It is now very much part of our everyday lives. As a consequence, the need for chemical analysis of the environment continues to grow.

This book is a thorough revision and expansion of the ACOL text 'Environmental Analysis' which was first published in 1994. It is an introduction into how, sometimes familiar, at other times less familiar, chemical analytical techniques are applied to the environment. A knowledge of basic analytical techniques is thus assumed. This could have been acquired, for instance, in the first two years of an undergraduate programme in chemistry or a related discipline. For the more familiar techniques the emphasis of the book is on the application of the technique, rather than on description of the basic principles. Examples include titration, UV/visible spectrometry and gas chromatography. More specialized techniques which would not be found in more general chemistry textbooks are described in more detail in the text, along with their application(s). Examples of these would be ion chromatography and solid extraction methods. Little more than a background knowledge of the environment is assumed, although an interest to learn about the subject is essential. A glossary, presented at the end of the book, provides a description of some of the less familiar terms.

The original (ACOL) book was aimed largely at background monitoring of the environment. Current interest requires a much wider area of coverage, in particular in monitoring liquid and gaseous discharges and surveying areas of past pollution. In this present text there is a larger section on solid sampling and extraction and sections on analysis of contaminated land and landfill are also included. More emphasis is placed on source monitoring. There is an expansion of quality assurance and quality control and more detail on quantification of the techniques.

A number of techniques which were emerging during the preparation of the original book have now become acceptable as alternatives to more long-standing methods. This is particularly the case with solid sample preparation, where a number of automated techniques have been developed and are now finding use in high-throughput laboratories. The monitoring of metals in water has also been transformed in the intervening years with the widespread introduction of inductively coupled plasma mass spectrometry (ICP-MS) and developments in the sensitivity of ICP-optical emission spectrometry (ICP-OES). Interest in field methods continues to grow, particularly in the area of rapid assessment to minimize the number of samples taken to the laboratory for analysis. This has included developments in techniques unfamiliar to many chemists, such as immunoassay and X-ray fluorescence spectrometry.

The techniques discussed develop in complexity, starting with simple volumetric measurements for water quality and finishing with ultra-trace analysis. Chapter 1 introduces you to simple concepts needed in the study of the environment, to what we mean by the term 'pollution' and the role of analytical chemistry. Chapter 2 starts by discussing pollution dispersion, reconcentration and final degradation – important concepts to understand when setting up a monitoring scheme. This chapter then goes on to describe simple concepts about sampling and the subsequent analysis, the choice of laboratory or field analysis, and also introduces quality assurance and quality control.

The remaining six chapters, in turn, cover the analysis of water, solid and atmospheric samples. Where there is a choice of techniques available, the questions (SAQs and DQs) guide you into understanding why one specific technique is often preferable. One of the main themes of this book is to demonstrate how an understanding of the principles of the analytical techniques is vital for good analytical choice. Chapters 3 and 4 are devoted to water, while Chapter 5 is concerned with solids and the techniques used to extract pollutants for subsequent analysis. This is an area of great current interest due to concern over waste dumping and potential problems with the reuse of old industrial sites. Chapters 6 and 7 are concerned with sampling and analysis of gases and particulates in external atmospheres, buildings and flues (chimneys or exhausts). Many of the techniques may already be familiar to you in the laboratory, although you will often find in the instruments very novel applications. Chapter 8 is concerned with the special problems of ultra-trace analysis.

A book of this length can only be seen as an introduction to environmental analysis. A bibliography is provided to guide you into more specialized texts in the area and to where you can find the various standard methods. It also gives examples of current usage of the techniques.

I would like to thank many people for their help in the production of this book – in particular to Rose Reeve for her support and endurance during its preparation, and for producing the drawings used as a basis for the illustrations in Figures 3.2, 3.4 and 6.3, and in the Response to SAQ 2.2. Some of these drawings are based on scenes around our home in Durham. Thanks are also due to colleagues at the University of Sunderland, to staff at the Environment

Agency, Leeds (UK), to Peter Walsh (HSE) for the diagram provided in the Response to SAQ 6.8, to Shirley and Steven Forster, Dorothy Hardy and Colin Edwards and to my students for all that I have learnt from their questioning. I would also like to thank the University of Sunderland for permission to use the following figures from the ACOL 'Environmental Analysis' book: 1.1–1.3, 2.5–2.8, 3.5–3.8, 3.10–3.12, 3.14, 3.17–3.19, 4.4, 4.5, 4.9, 4.14, 4.15, 4.20, 4.21, 5.1–5.3, 6.1, 6.2, 6.4, 6.7–6.10, 6.12, 6.13, 6.15–6.17, 6.20, 8.2, 8.3, 8.6 and 8.7.

Finally, I hope that this book will be a true introduction to the subject and will lead you into further study in the exciting area of environmental analysis.

Roger Reeve
University of Sunderland, UK

Acronyms, Abbreviations and Symbols

AAS	atomic absorption spectrometry
AC	alternating current
amu	atomic mass unit (dalton)
ASTM	American Society for Testing and Materials (USA)
ASV	anodic stripping voltammetry
BOD	biochemical oxygen demand
BSI	British Standards Institute (UK)
BTEX	benzene–toluene–ethylbenzene–xylene(s)
CFC	chlorinated fluorocarbon
COD	chemical oxygen demand
Da	dalton (atomic mass unit)
DC	direct current
DDT	p, p'-dichlorodiphenyltrichloroethane
DOAS	differential optical absorption spectrometry
EA	Environment Agency (UK)
EDTA	ethylenediaminetetraacetic acid
EEC	European Economic Community
ELISA	enzyme-linked immunosorbent assay
emf	electromotive force
EPA	Environmental Protection Agency (USA)
EU	European Union
eV	electronvolt
FTIR	Fourier-transform infrared
GC	gas chromatography
GFAAS	graphite furnace atomic absorption spectrometry
GL	guide level (EU)

GLP	Good Laboratory Practice (OECD)
GQA	General Quality Assessment (UK)
HCFC	hydrochlorofluorocarbon
HFC	hydrofluorocarbon
HMIP	Her Majesty's Inspectorate of Pollution (UK)
HMSO	Her Majesty's Stationary Office (UK)
HPLC	high performance liquid chromatography
HSE	Health and Safety Executive (UK)
IC	ion chromatography
ICP	inductively coupled plasma
i.d.	internal diameter
IR	infrared
ISO	International Organization for Standardization
J	joule
LC	liquid chromatography
LIDAR	light detection and ranging
MAC	maximum admissible concentration (EU)
MDHS	Methods for the Determination of Hazardous Substances (UK)
MEL	maximum exposure limit (UK)
MS	mass spectrometry
NAMAS	National Accreditation Management Service (UK)
NAQS	National Air Quality Standard (USA)
NIOSH	National Institute of Occupational Safety and Health (USA)
NO_x	$NO + NO_2$
NTIS	National Technical Information Service (USA)
ODS	octadecylsilane
OECD	Organization for Economic Co-operation and Development
OES	optical emission spectrometry; occupational exposure standard (UK)
PAH	polynuclear aromatic hydrocarbon
PAN	peroxyacetyl nitrate
PCB	polychlorinated biphenyl
PCDD	polychlorinated dibenzo-*p*-dioxin
PCDF	polychlorinated dibenzofuran
PM_{10}	particle with aerodynamic diameter less than 10 μm
ppb	parts per billion (1 part in 10^9)
ppm	parts per million
PTFE	polytetrafluoroethylene
PVC	poly(vinyl chloride)
RF	radiofrequency
rms	root mean square
sd	standard deviation
SFC	supercritical fluid chromatography
SFE	supercritical fluid extraction

SI (units)	Système International (d'Unitès) (International System of Units)
SO_x	$SO_2 + SO_3$
SPE	solid-phase extraction
TDLAS	tuneable diode laser absorption spectroscopy
TDS	total dissolved solids
TEOM	tapered element oscillating microbalance
TEQ	toxic equivalent concentration
TOC	total organic carbon
TPH	total petroleum hydrocarbon
TWA	time-weighted average
UNEP	United Nations Environmental Programme
UV	ultraviolet
V	volt
VOC	volatile organic compound
W	watt
XRF	X-ray fluorescence

c	speed of light; concentration
e	electronic charge (charge on an electron)
E	energy; electric field strength
f	(linear) frequency
I	electric current
m	mass
m/z	mass/charge ratio (mass spectrometry)
$M_r(X)$	relative molecular mass (of X)
p	pressure
Q	electric charge (quantity of electricity)
R	molar gas constant; resistance
t	time; Student factor
T	thermodynamic temperature
V	electric potential
z	ionic charge
Z	atomic number
λ	wavelength
ν	frequency (of radiation)
σ	standard deviation
σ^2	variance

About the Author

Roger N. Reeve, B.Sc., M.A., Ph.D.

Roger Reeve took his first degree in Natural Science at Oriel College, Oxford and went on to the University of Durham to obtain a doctorate in Inorganic Chemistry. He then spent several years in the research and development department of a process plant manufacturing company which specialized in pollution control equipment for large-scale industrial processes. Much of this work dealt with gaseous pollutants. It was here that he developed his scientific interest in chemical analysis and the environment with the realization that analysis can extend far outside the laboratory. His work included one of the earliest applications of reversed phase ion-pair liquid chromatography to the separation of inorganic ions. He then returned to academic life at the University of Bradford and, from 1985, at the University of Sunderland, where he is now Senior Lecturer in Analytical and Inorganic Chemistry in the Institute of Pharmacy, Chemistry and Biomedical Sciences. His research interests are within the Pharmaceutical and Environmental Analysis Group of the Institute, including the development of immunoassays for atmospheric pollutants. As well as environmental analysis, he teaches environmental and inorganic chemistry.

Chapter 1
Introduction

Learning Objectives

- To explain what is meant by the term 'environment'.
- To identify reasons for concern over the current and future quality of the environment.
- To appreciate the diversity of pollution.
- To evaluate the role of chemical analysis in dealing with these problems.

1.1 The Environment

We live in a world where the environment is of major concern. In our newspapers we read of governments attempting to find agreement over global environmental problems. We can use 'green' fuel in our transport, shop for 'environmentally friendly' products and recycle much of our waste. However, what do we mean by our **environment**? Are we referring here to:

The place where we live or work?
The atmosphere which we breathe and the water which we drink?
Unspoilt areas of the world which could soon be ruined?
Parts of the atmosphere which shield us from harmful radiation?

The environment must include all of these areas and anywhere else which could affect the well-being of living organisms. Concern must extend over any process which would affect this well-being, whether it is physical (e.g. global warming and climate change), chemical (e.g. ozone layer depletion) or biological (e.g. destruction of rain forests).

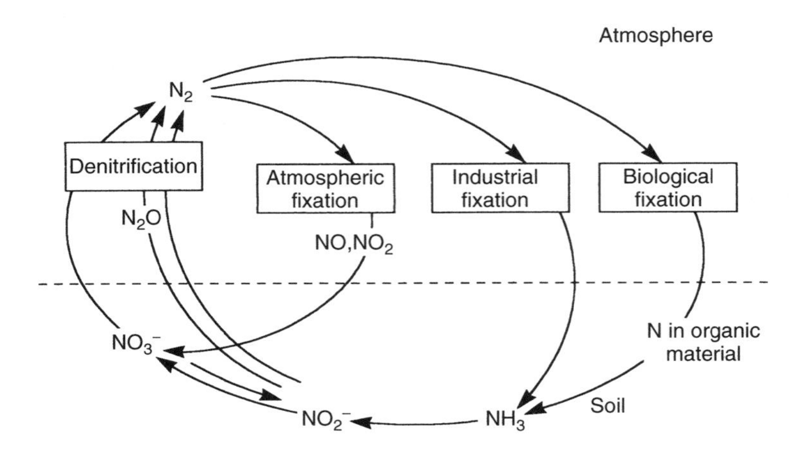

Figure 1.1 Illustration of a simplified nitrogen cycle.

Anyone who has more than a passing interest in the environment has to learn and understand a very broad range of subjects. The purpose of this introduction is first of all to show how analytical chemistry fits into this broad spectrum, and later to demonstrate how it is an essential part of any scientific study of the environment and its problems. The book then goes on to discuss how analytical chemistry is applied to the three spheres of the environment, namely water, land and atmosphere.

In order to understand the environment, we must first realize that it is never static. Physical forces continuously change the surface of the earth through weather, the action of waves and natural phenomena, such as volcanoes. At the same time, they release gases, vapour and dust into the atmosphere. These can return to the land or sea a great distance away from their sources. Chemical reactions high up in the atmosphere continuously produce ozone which protects us from harmful ultraviolet radiation from the sun. Living organisms also play a dynamic role through respiration, excretion, and ultimately, death and decay, thus recycling their constituent elements through the environment. This is illustrated by the well-known nitrogen cycle (Figure 1.1). There are similar cycles for all elements which are used by living organisms.

1.2 Reasons for Concern

The current interest in the environment stems from the concern that the natural processes are being disrupted by people to such an extent that the quality of life, or even life itself, is being threatened.

Many indicators would suggest that the world is at a crisis point; for instance, the rapid population growth of the world, as shown in Figure 1.2, and the

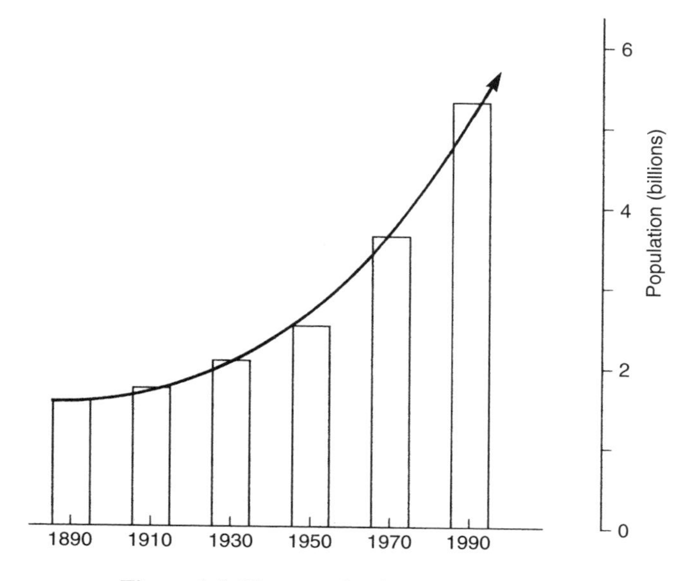

Figure 1.2 The growth of world population.

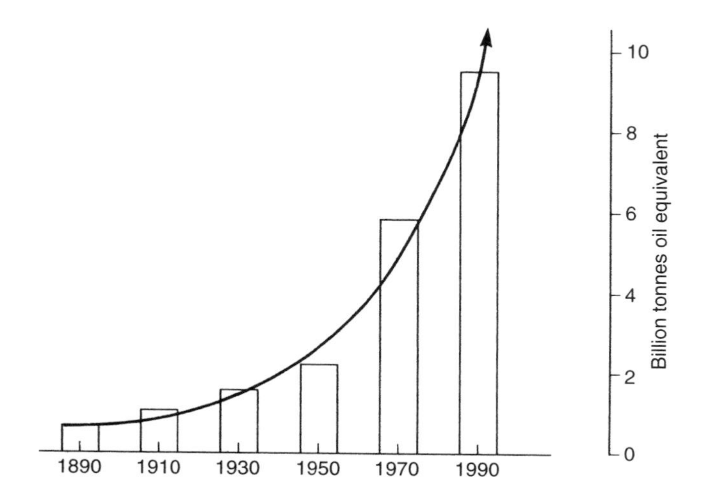

Figure 1.3 The growth of energy consumption.

consequential growth in energy consumption shown in Figure 1.3. Not only will the earth be depleted of its resources, with the inevitable environmental damage that will result, but there will almost certainly be a parallel increase in waste produced and in pollution of the earth. The increase in production of carbon dioxide follows an almost identical curve to the energy consumption increase.

This concern has become heightened by a greater awareness of problems than in previous ages, due to greater ease of communication, which bring news from distant parts of the world. It seems ironic that the greater prosperity of the developed world, giving sufficient leisure time for concern over global problems, but also giving increased resource consumption, is currently a large contributing factor to the problems themselves.

1.2.1 Today's World

The type of discussion above can lead to a pessimistic view of the future. However, there has been much national and international legislation leading to the control of pollution, and the ordinary person in the street can immediately see the benefits of taking a greater concern for the environment. The choking sulfurous fog which used to engulf London on winter days is now only found in history books. The lower reaches of the River Thames were once dead but now it is one of the cleanest in Europe, with at least 115 different species of fish. Care of the environment is on everyone's lips and in their lifestyle. There are few people who will never have heard of the potential problems of increased greenhouse gas emissions. Legislation is continuously being introduced to improve our environment. In many countries, we have moved to the stage where concern for the environment is an integral part of everyday life.

1.2.2 Past and Current Crimes

Some of the concern today is centred on problems inherited from less enlightened ages which will be with us for many years to come. Examples include spoil heaps from mining operations, contaminated land from previous industrial sites, and pesticides which are now banned but have such a long lifetime in the environment that they will continue to pollute for many decades. Current concerns include emissions from our automobiles, waste production, production of toxic particulate matter from combustion and incineration processes, use of pesticides which build up in the food chain and the use of inorganic fertilizers in agriculture. Although more environmentally friendly methods for power production are being introduced, there is still a large-scale reliance on fossil fuel for energy production with its inevitable production of carbon dioxide.

1.3 Pollution

All of us have concepts of what pollution is but have you considered how it may be defined?

DQ 1.1

What you would consider to be a definition of pollution?

Answer

The following definition is from the Organization of Economic Co-operation and Development:

'Pollution means the introduction by man, directly or indirectly, of substances or energy into the environment resulting in deleterious effects of such a nature as to endanger human health, harm living resources or interfere with amenities or other legitimate use of the environment.'

Before we concentrate on the chemical aspects of pollution, it is worth remembering that this is not the only form of pollution. Noise is an example of physical pollution. Simply adding water to a river at a different temperature to the ambient can effect life in the river. This is a form of thermal pollution. Pollution is, however, often associated with the introduction of chemical compounds into the environment. Popular opinion usually sees these as unnatural (and therefore harmful) substances. Perhaps one of the best known recent examples was the concern over the emission of chlorofluorocarbons (CFCs). These have been used in aerosol sprays and other applications. They are linked with the depletion of ozone in the stratosphere, which could lead to an increase in the intensity of harmful ultraviolet radiation from the sun reaching the earth's surface and increasing the incidence of skin cancer. Although the production of CFCs themselves is now banned in developed countries, the existing CFCs will take many years to be removed from the atmosphere and related ozone-depleting compounds (e.g. hydrochlorofluorocarbons, (HCFCs)) are still being manufactured. The effects on the ozone layer will therefore remain for many decades.

More frequently, problems occur by the release of substances into the environment which are naturally present, with the problem arising simply from an increase in concentration above the 'natural' levels. Carbon dioxide is a natural component of the atmosphere produced by the respiration of living organisms. The potential problem of global warming is primarily associated with an increase in its concentration in the atmosphere as a result of fuel combustion, together with a decrease in the world's forests which recycle the carbon. Increasing concentrations of a number of other naturally occurring gases, such as methane and nitrous oxide, add to the problem. Nitrates occur naturally as part of the constant cycling of nitrogen in the environment (see Figure 1.1). The over-use of fertilizers can, however, produce a build-up of nitrate in water courses which leads, first of all, to excessive plant growth, but ultimately to the death of all living species in the water. The process is known as *eutrophication*. Apart from nitrogen itself, all of these species in the nitrogen cycle have been shown to exhibit environmental problems if their concentration increases greatly above the 'natural' level in water or in the atmosphere. This is summarized in Table 1.1

You should be able to think of many pollution examples of your own. Try grouping the problems into different categories, for instance, whether the pollution is a global problem (e.g. ozone-depletion) or a more local issue (e.g. waste dumping). When you read the next chapter, which deals with the transport of

Table 1.1 Examples of problems caused by excessive concentrations of nitrogen species

Species	Problem
N_2O	Contributes to the greenhouse effect and is a potential ozone-depleter
NH_3	Highly poisonous to fish, particularly in its non-protonated form
NO_2^-	Highly poisonous in water to animals
NO_3^-	Contributes to eutrophication (excessive plant growth) in watercourses; associated with 'blue-baby syndrome' which can cause fatalities in infants

pollutants, you may find that you change your mind about some of the problems. Lead pollution, which has been associated with the retardation of intellectual development in children, is normally thought to be a highly localized problem. Increased lead concentrations in the environment, largely from the use of leaded petrol in cars, can be detected hundreds of kilometres from likely sources.

DQ 1.2

If a pollutant is discharged into the environment, what causes the effect on individual living organisms:

- the total amount discharged;
- its concentration in the environment?

Answer

It is the ***concentration*** *which is of concern with respect to individual living organisms.*

This statement may seem surprising but consider the following facts. All compounds are toxic at high enough concentrations. Even something apparently as innocuous as sodium chloride has adverse effects when present in high concentration. For example, you cannot drink more than a small quantity of sea water without being made ill. Some metals, which are necessary for plant growth when found in small concentrations in the soil, would kill the plant life when found in larger concentrations on, let us say, a waste dump. These include elements such as chromium, cobalt and manganese, and are often known as 'essential' elements.

Of course, if we are considering the effect of a particular pollutant on the global environment, we would have to consider the total quantity emitted. Excessive amounts would ultimately increase the background concentration, as is the case with carbon dioxide emissions.

It would then appear, that in order to limit the adverse effect of a particular ion or compound, it is necessary to ensure that the concentration in water or in the atmosphere is maintained below a pre-determined 'safe' level. As will be shown in the next section, the establishment of such levels is fraught with difficulty. Nonetheless, much of the world's environmental legislation is drafted in terms of specifying maximum concentration of ions and compounds (Table 1.2).

Table 1.2 Extract from European Community Directive 80/778/EEC relating to the quality of water intended for human consumption – parameters concerning substances undesirable in excessive amounts[a]. Reproduced by permission of the Official Journal of the European Communities

Parameter	Expression of the results[a]	Guide level (GL)	Maximum admissible concentration (MAC)	Comments
20 Nitrates	NO_3 (mg l^{-1})	25	50	—
21 Nitrites	NO_2 (mg l^{-1})	—	0.1	—
22 Ammonium	NH_4 (mg l^{-1})	0.05	0.5	—
23 Kjeldahl nitrogen (excluding N in NO_2 and NO_3)	N (mg l^{-1})	—	1	—
24 ($KMnO_4$) oxidizability	O_2 (mg l^{-1})	2	5	Measured when heated in acid medium
25 Total organic carbon (TOC)	C (mg l^{-1})			The reason for any increase in the usual concentration must be investigated
26 Hydrogen sulfide	S (μg l^{-1})	—	Undetectable organoleptically	—
27 Substances extractable in chloroform	Dry residue (mg l^{-1})	0.1	—	—
28 Dissolved or emulsified hydrocarbons (after extraction by petroleum ether); mineral oils	μg l^{-1}	—	10	—
29 Phenols (phenol index)	C_6H_5OH (μg l^{-1})	—	0.5	Excluding natural phenols which do not react with chlorine
30 Boron	B (μg l^{-1})	1000	—	—
31 Surfactants (reacting with methylene blue)	Lauryl sulfate (μg l^{-1})	—	200	—

[a]Certain of these substances may even be toxic when present in very substantial quantities.

DQ 1.3

What are the maximum concentrations that the substances listed in Table 1.2 can be considered to be acceptable in drinking water?

Answer

These, of course, vary from substance to substance, but you should have noted that most of the maximum admissible concentrations are expressed in units of mg l^{-1} (sometimes called parts per million (ppm)), whereas others are expressed as μg l^{-1} (or parts per billion (ppb)).

SAQ 1.1

How would you see the following situations as contributing to pollution problems?

1. An increase in the developed world's population.
2. Volcanic emissions.
3. Production of methane by cows, as part of their natural digestion.
4. Excessive quantities of nitrate fertilizers used in farming.

1.4 The Necessity of Chemical Analysis

If you were performing a simple pollution monitoring exercise, it is evident that a detailed analysis of pollution levels would be an essential part. Let us now consider a complete control programme and look in detail at what stages chemical analysis would be necessary.

DQ 1.4

List what steps you think would be necessary for a national government or international agency to control a potential pollution problem, starting from the initial recognition. At what stages would chemical analysis be involved?

Answer

1. Recognition of the Problem

This would appear to be an obvious statement until you consider how recently many pollution problems have become recognized. The term 'acid rain' originally referred to localized effects of sulfur oxides (SO_2 and SO_3) produced from coal combustion and was introduced in the 19th century. Trans-national problems, such as may arise from the transport of the gases from the power stations in the north of England to Scandinavia, have only been recognized in the last three decades. The contribution of other chemical compounds, such as nitrogen oxides (NO and NO_2), to acid rain was only acknowledged several years later. Alternatives to the ozone-depleting CFCs were introduced in the late 1980s and early 1990s. These included hydrofluorocarbons (HFCs) which have no ozone-depleting potential. There was little regard originally taken of their large greenhouse-warming effect. Currently, there is much concern over endocrine disruptors, known in the popular press by terms such as 'gender benders' or 'sex-change chemicals', which have recently been shown to effect the early stages of foetal development in some species. This leads to mixed sexual characteristics, usually seen as the feminization of males. Such compounds are widespread in the environment. Some have long been known to have environmental effects (e.g. polychlorinated biphenyls and the pesticide DDT), while others had been previously considered completely benign (e.g. phthalate esters which are used as plasticizers in PVC materials).

2. Monitoring to Determine the Extent of the Problem
As we have already seen, this may either involve analysis of a compound not naturally found in the environment, or determination of the increase in concentration of a compound above the 'natural' level. The determination of 'natural' levels could itself involve a substantial monitoring exercise since these levels may vary greatly with location and season. Large quantities of waste materials have been produced for many centuries, and it may even be a difficult task to assess what an unpolluted environment is. For example, it has been discovered that the highly toxic and potentially carcinogenic compounds commonly referred to as 'dioxins', which were originally assumed to be completely anthropogenic (man-made), occur naturally at trace levels.

3. Determination of Control Procedures
Determination of the most appropriate method should involve testing the options with suitable analytical monitoring. Possibilities include technological methods, such as the use of flue gas desulfurization processes to lower sulfur oxide emissions from coal-fired power stations, and socially orientated methods, such as the promotion of the use of public rather than private transport to reduce vehicle emissions.

4. Legislation to Ensure the Control Procedures are Implemented
Few pollution control methods are taken up without the backing of national or international legislation. As shown in Table 1.2, this legislation is very often drafted in terms of analytical concentrations.

5. Monitoring to Ensure the Problem has been Controlled
A large proportion of current monitoring is to ensure compliance with legislation. This may range from national programmes to confirm air and

water quality to local monitoring of discharges from industries and to the yearly checking of emissions from individual automobiles. Monitoring also provides scientific evidence for possible further developments in legislation.

Have you noticed the cyclical nature of the process which includes monitoring to show that a problem exists, reduction of the problem by control procedures, and monitoring to confirm that the problem has been reduced, with the final stage leading back to the start for improvement in the control procedures?

You should also have noticed that chemical analysis is a necessary component of *almost all* of the stages!

SAQ 1.2

Consider a factory producing a liquid discharge, consisting partly of side products of the process and partly of contaminants present in the starting materials. What analytical monitoring programme would be useful to assess and control the effluent?

Summary

This introduction answers the question of what is meant by the terms 'environment' and 'pollution'. Pollutants are often materials which are naturally present in the environment, with their adverse effects being caused by concentrations higher than those which would be expected from natural causes. A study of pollution would then involve a large amount of quantitative chemical analysis. Analytical chemistry is also involved in devising pollution control procedures, in drafting legislation and in monitoring the effect of any control procedure. In fact, analytical chemistry is a necessary component in almost all aspects of scientific investigations of the environment, the problems caused by mankind and their possible solution.

Chapter 2

Transport of Pollutants in the Environment and Approaches to their Analysis

Learning Objectives

- To predict the possible movements of a pollutant in the environment.
- To suggest sampling locations where high-molecular-mass organic compounds and metals may accumulate.
- To define what is meant by the terms 'critical path' and 'critical group'.
- To introduce sampling and sample variability.
- To understand the range of methods needed for subsequent chemical analysis.
- To introduce quality assurance.

2.1 Introduction

We have learnt how the environmental effects of compounds are dependent on their concentration and also that the environment is not static. Materials are constantly being transported between the three spheres of the environment – the atmosphere, the hydrosphere and the lithosphere (the earth's crust). At each stage of the transportation, the concentration of the compounds will be altered either by phase transfer, dilution or, surprisingly, reconcentration. Before discussing analytical methods, we need to understand these processes so that we can:

- predict where large concentrations of the pollutant are likely to occur;
- assess the significance of measured concentrations of pollutants in different regions of the environment.

For this we need to discuss the chemical and physical properties of the pollutant. This will also help us to identify species which may be of particular concern, and to understand why, of the many thousands of ions and compounds regularly discharged into the environment, particular concern often centres on just a few classes.

2.2 Sources, Dispersal, Reconcentration and Degradation

Virtually every form of human activity is a potential source of pollution. The popular concept of industrial discharge being the primary source of all pollution is misguided. It is just one example of a **point** source, i.e. a discharge which can be readily identified and located. Discharges from sewage works provide a second example. In some areas these are the major source of aquatic pollution. Sometimes, however, it is not possible to identify the precise discharge point. This can occur where the pollution originates from land masses. Examples include the run-off of nitrate salts into watercourses after fertilizer application and the emission of methane from land-fill sites into the atmosphere. These are examples of **diffuse** sources.

Both water and the atmosphere are major routes for the dispersal of compounds. What comes as a surprise are the pathways by which some of the compounds disperse. It is very easy, for instance, for solid particulate material to be dispersed long distances via the atmosphere. There has been, for example, an approximately equal quantity of lead entering the North Sea off the coast of Britain from atmospheric particulates as from rivers or the dumping of solid waste. To illustrate this, a typical transport scheme for a metal (lead) is shown in Figure 2.1.

Equally surprising are the dispersal routes of 'water-insoluble' solid organic compounds. No material is completely insoluble in water. For instance, the solubility in water of the petroleum component, isooctane (2,2,4-trimethylpentane), is as high as 2.4 mg l^{-1}. Watercourses provide a significant dispersal route for such compounds.

The significant vapour pressure of organic solids is also often forgotten. Consider how readily a solid organic compound such as naphthalene, as used in mothballs, volatilizes. In these cases, transportation through the atmosphere is partly in the solid phase and partly in the vapour phase. If you wish to monitor the concentration of these materials in the atmosphere, you not only have to analyse the suspended particulate material but also the gaseous fraction.

The atmosphere also provides a dispersal route for volatile organic compounds. Hydrocarbons will be quickly degraded but will contribute to localized pollution

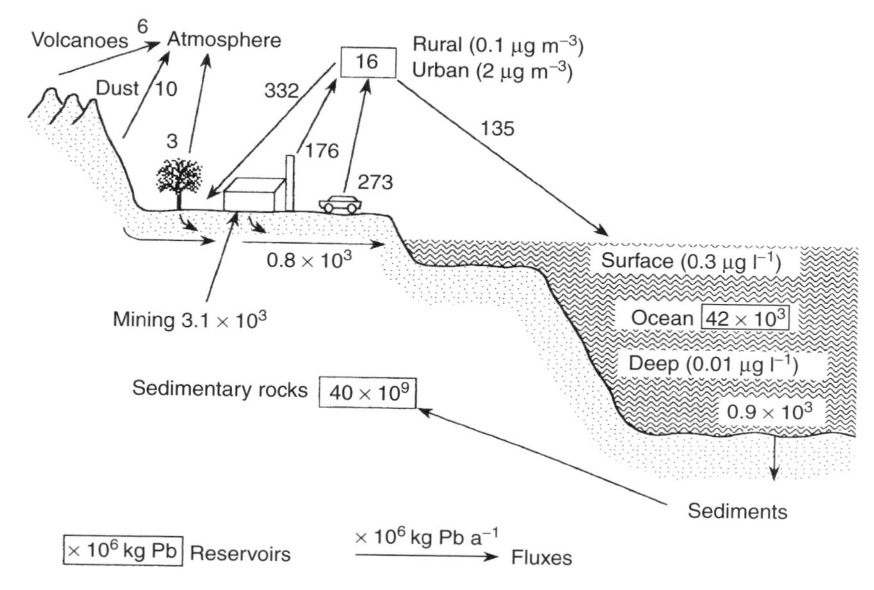

Figure 2.1 Transport of lead in the environment; concentrations are given in parentheses. Reproduced with the permission of Nelson Thornes Ltd from Environmental Chemistry 3rd Edition ISBN No 0 7514 04837 first published in 1998.

in the form of photochemical smog. If the compound is stable, or is only slowly degraded, in the lower atmosphere, as is the case with many chlorine- or bromine-containing compounds, some may eventually reach the stratosphere (the portion of atmosphere at an altitude of 10–50 km). Decomposition, promoted by the intensity of low-wavelength radiation at this altitude, initiates a series of chemical reactions which deplete the protective layer of ozone.

Distances which are travelled by pollutants in the atmosphere may be as long as hundreds or thousands of kilometres. The movement of sulfur oxides has been studied over distances covering the whole of Europe, and when Mount St. Helens volcano erupted in the USA, the particulate material which was discharged resulted in the production of vivid sunsets several thousand kilometres away.

Dispersal of a pollutant in water or in the atmosphere will inevitably lead to a dilution of the pollutant. As we have seen that the effect of a chemical compound in the environment can be related directly to its concentration, you may think that the dispersal process will simply spread out the pollutant such that it could have little effect away from the source. This would especially be the case when we consider that most forms of pollution are eventually broken down by microbial attack, photochemical or other degradation, and so there would be little chance of the concentration building up to toxic levels. Indeed the phrase 'Dilution is the solution to pollution' was often heard in the early days of environmental concern.

DQ 2.1

What factors do you think this statement does not take into account?

Answer

(a) The possibility that some pollutants can reconcentrate at particular locations or within organisms remote from the original source.

(b) The non-degradation or slow degradation of some pollutants so that there is a gradual concentration buildup in the environment at large.

(c) Contamination of large areas before sufficient dilution has taken place.

Examples may be given for all these cases, as follows:

(a) Toxic metals, such as cadmium, may be found in the organs of shellfish in concentrations up to 2 million times greater than in the surrounding water (Table 2.1).

(b) The major constituent of the pesticide DDT (p,p'-dichlorodiphenyltrichloroethane) is now a universal contaminant due to its widespread use over several decades and its slow degradation. There is little organic material on the earth which does not contain traces of this at the ng 1^{-1} level or greater concentration.

(c) Dilution does not take into account localized pollution effects which may occur around discharge pipes or chimneys before dispersion occurs. One of the observed effects of pollution by endocrine disruptors is the 'feminization' of male fish. This particularly occurs close to sewage outfalls where several of the compounds first enter the environment.

The effects of pollution have also been often underestimated in the past. The discharge of sulfur dioxide in gases from tall chimneys was, until recently, seen as an adequate method for its dispersal. The potential problem of 'acid rain' was not considered.

Table 2.1 Examples of metal enrichment in shellfish relative to the surrounding water

Metal	Relative concentration in shellfish[a]
Cadmium	2 260 000
Chromium	200 000
Iron	291 500
Lead	291 500
Manganese	55 500
Molybdenum	90
Nickel	12 000

[a] Water = 1.

The following sections will discuss two major categories of pollutants which have caused environmental concern due to their ability to reconcentrate (accumulate) in specific areas and within living organisms. These provide good examples of how a knowledge of the transport of pollutants can be used to determine suitable sampling locations where high concentrations may be expected.

> **SAQ 2.1**
>
> What general physical and chemical properties would you expect in a compound which has become a global pollution problem?

2.3 Transport and Reconcentration of Neutral Organic Compounds

Compounds in this category which readily reconcentrate and are of global concern are usually of low volatility and high relative molecular mass ($M_r > 200$). They often contain chlorine atoms within the molecule. Some typical compounds are shown in Figure 2.2.

Compounds of lower relative molecular mass may produce severe local atmospheric problems. Hydrocarbon emissions from automobiles are currently of concern due to their contribution to the photochemical smog which affects large cities throughout the world. These effects occur where the climate and geographical conditions permit high atmospheric concentrations to build up with little dispersal. However, unless the compounds are particularly stable to decomposition within the atmosphere (as is the case with chlorofluorocarbons), or are

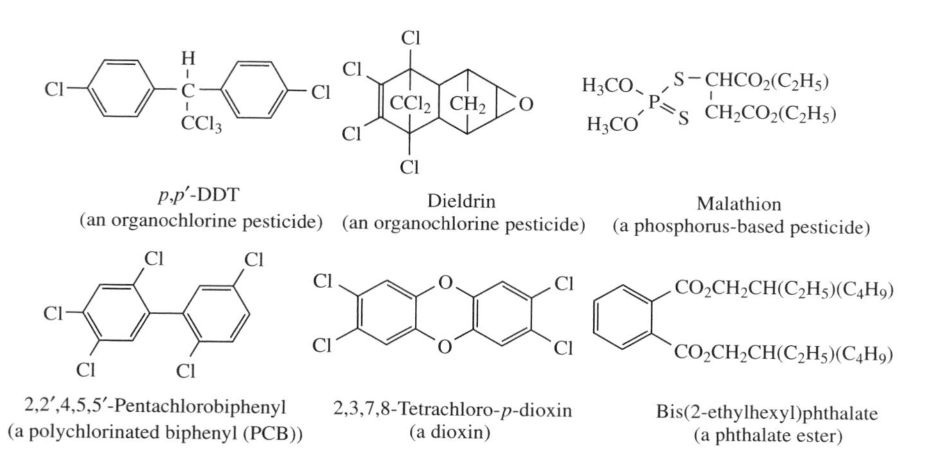

Figure 2.2 Some examples of neutral organic compounds of environmental concern.

discharged in such great quantities that they can build up globally (as is the case with methane), they will remain local, rather than global, pollutants.

We will now discuss the mechanisms by which organic compounds can reconcentrate within organisms, and will discover one of the reasons why it is the compounds of higher relative molecular mass that are of greatest concern.

2.3.1 Bioconcentration

Unless organic compounds contain polar groups such as $-OH$, or $-NH_2$, or are ionic, they will have low solubility in water. Within related groups of compounds, the solubility decreases with increasing molecular mass. As the solubility in water decreases, the solubility in organic solvents increases (Figure 2.3). This increase in solubility is equally true if we consider solubility in fatty tissues in fish and aquatic mammals rather than solubility in laboratory solvents. Any dissolved organic material will readily transfer into fatty tissue, particularly that found in organs in closest contact with aqueous fluids, e.g. kidneys.

DQ 2.2

What rule can you deduce concerning the solubility of a compound in water, and its ability to accumulate in organisms?

Answer

We arrive at a very unexpected general rule that ***the lower the solubility of an organic compound in water, than the greater is its ability***

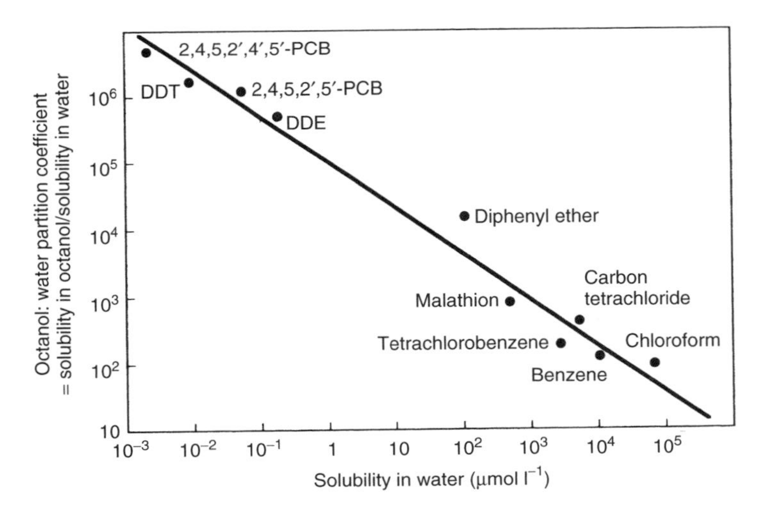

Figure 2.3 Partition coefficients versus aqueous solubilities of environmentally significant organic compounds. Reprinted with permission from Chiou, C.T., Freed, V.H., Schnedding, D.W. and Kohnert, R.L., *Environ. Sci. Technol.*, **11**, 475–478 (1977). Copyright (1977) American Chemical Society.

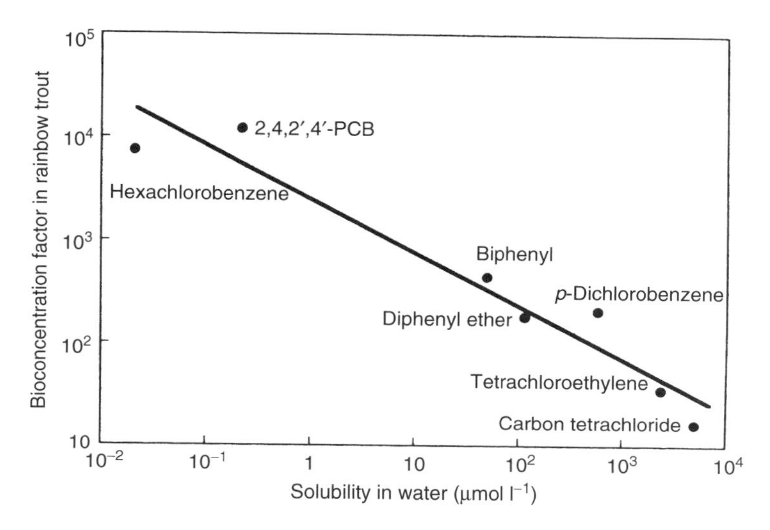

Figure 2.4 Bioconcentration factors versus aqueous solubilities of environmentally significant organic chemicals in rainbow trout. Reprinted with permission from Chiou, C.T., Freed, V.H., Schnedding, D.W. and Kohnert, R.L., *Environ. Sci. Technol.*, **11**, 475–478 (1977). Copyright (1977) American Chemical Society.

> ***to accumulate in fatty tissues and the greater is the potential for toxic effect.*** *In addition, because the solubility in water decreases with increasing molecular mass for related groups of compounds, we could also deduce that higher-molecular-mass compounds will pose greater aquatic environmental problems than compounds of lower molecular mass.*

The rule is illustrated in Figure 2.4, where the ability to accumulate in an organism is measured by the bioconcentration factor, as defined in the following equation:

$$\text{Bioconcentration factor} = \frac{\text{Concentration of a compound in an organism}}{\text{Concentration in surrounding water}} \quad (2.1)$$

2.3.2 Accumulation in Sediments

This is also related to the low solubility of high-molecular-mass organic compounds in water, together with the hydrophobicity of organic compounds not containing polar groups. Undissolved or precipitated organic material in water will adhere to any available solid. The larger the solid surface area, then the greater will be its ability to adsorb the compound. Suitable material is found in sediments. This is particularly true within estuaries where there are often discharges from major industries and fine sediment is in abundance. It is often the case (as may be expected from surface area considerations) that the smaller the particle size, then the greater is the accumulation of organic compounds in the sediment. These organics may then be ingested by organisms which feed by filtration of sediments (e.g. mussels, scallops, etc.) or, if the solid is sufficiently fine to be held in suspension, by 'bottom-dwelling' fish.

2.3.3 Biomagnification

Animals obtain their food by feeding on other plants or animals. Food chains can be built up where one species is dependent for survival on the consumption of the previous species. If a pollutant is present in the first organism, then as we proceed down the food chain there will be an increase in concentration in each subsequent species. This is illustrated in Figure 2.5.

Although the concept of such food chains is much simplified from the situation which occurs in nature (few species have just one source of food), it does provide an explanation for why the greatest concentration of pollutants is found in birds of prey at the end of the food chain, rather than in organisms in closest contact with the pollutant when originally dispersed.

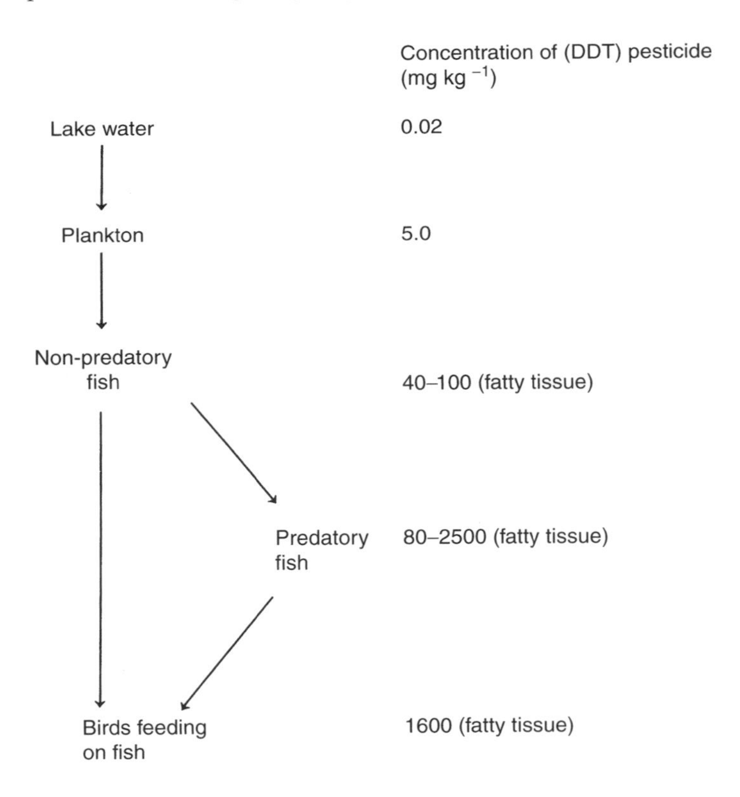

Figure 2.5 Illustration of a typical food chain.

2.3.4 Degradation

Even if a compound has a tendency to transfer into organisms by the routes described, it will not build up in concentration within the organism if it is rapidly metabolized. Compounds will break down until a molecule is produced with sufficient water solubility to be excreted. The solubility may be due either to polar groups being attached to the molecule or to its low relative molecular mass.

The rate of metabolism is highly dependent on the structure of the molecule. One of the reasons why so many organic compounds of environmental concern contain chlorine atoms is due to the slow metabolism of many of these compounds.

If we take *p,p′*-DDT as an example, the metabolism of this compound occurs in two stages, as shown in Figure 2.6. The first stage is rapid, and normally takes only a few days for completion, while the second stage is extremely slow, taking many months in some species. It is, in fact, the first degradation product which is often the predominant species in environmental samples. A minor component of

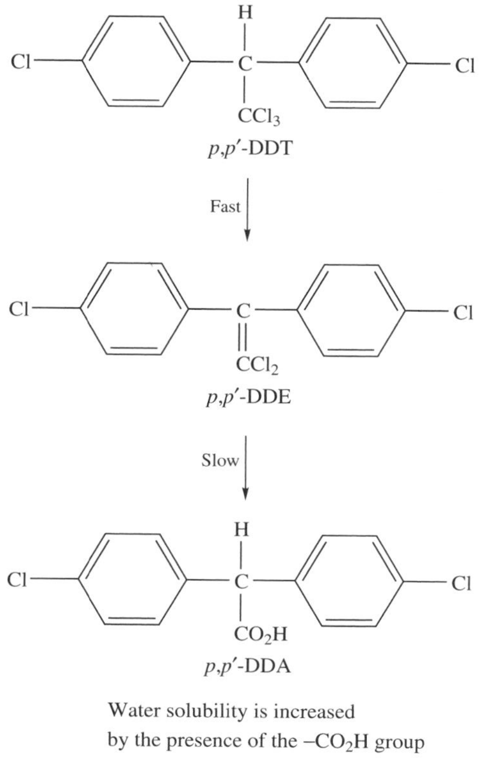

Figure 2.6 Metabolism of *p,p′*-DDT.

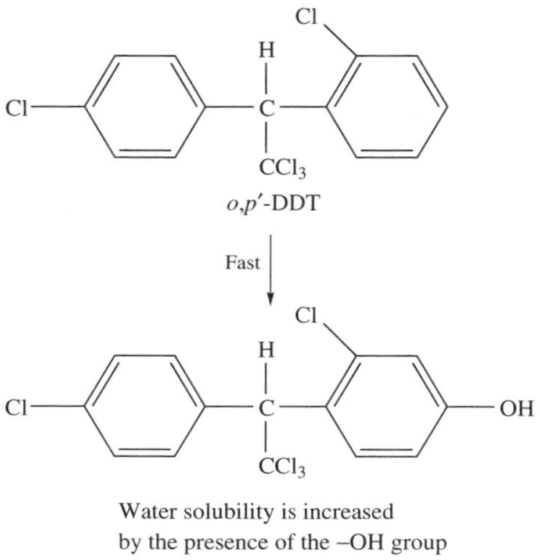

Figure 2.7 Metabolism of *o,p′*-DDT.

commercial DDT is the *o,p′*-isomer. This is metabolized rapidly by the reaction shown in Figure 2.7, and so does not accumulate significantly in organisms.

SAQ 2.2

Consider a pesticide such as DDT being sprayed on to a field from an aeroplane. Sketch routes by which the pesticide may disperse from the area of application.

2.4 Transport and Reconcentration of Metal Ions

We were able to discuss the movement of neutral organic compounds in simple terms because often very little chemical change occurs to the compounds during transportation through the environment and the initial degradation products frequently have similar physical and chemical properties to the parent compound. Unfortunately this is not the case with many of the metals of environmental concern. Their reaction products often have vastly different chemical and physical properties.

The metals which are of most environmental concern are first transition series and post transition metals (Figure 2.8), many of which are in widespread use in industry. Often, the non-specific term 'heavy metals' is used for three of the metals, namely lead, cadmium and mercury. These have large bioconcentration factors in marine organisms (look at the values for lead and cadmium in Table 2.1), are highly toxic and, unlike many of the transition elements, have no known natural biological functions.

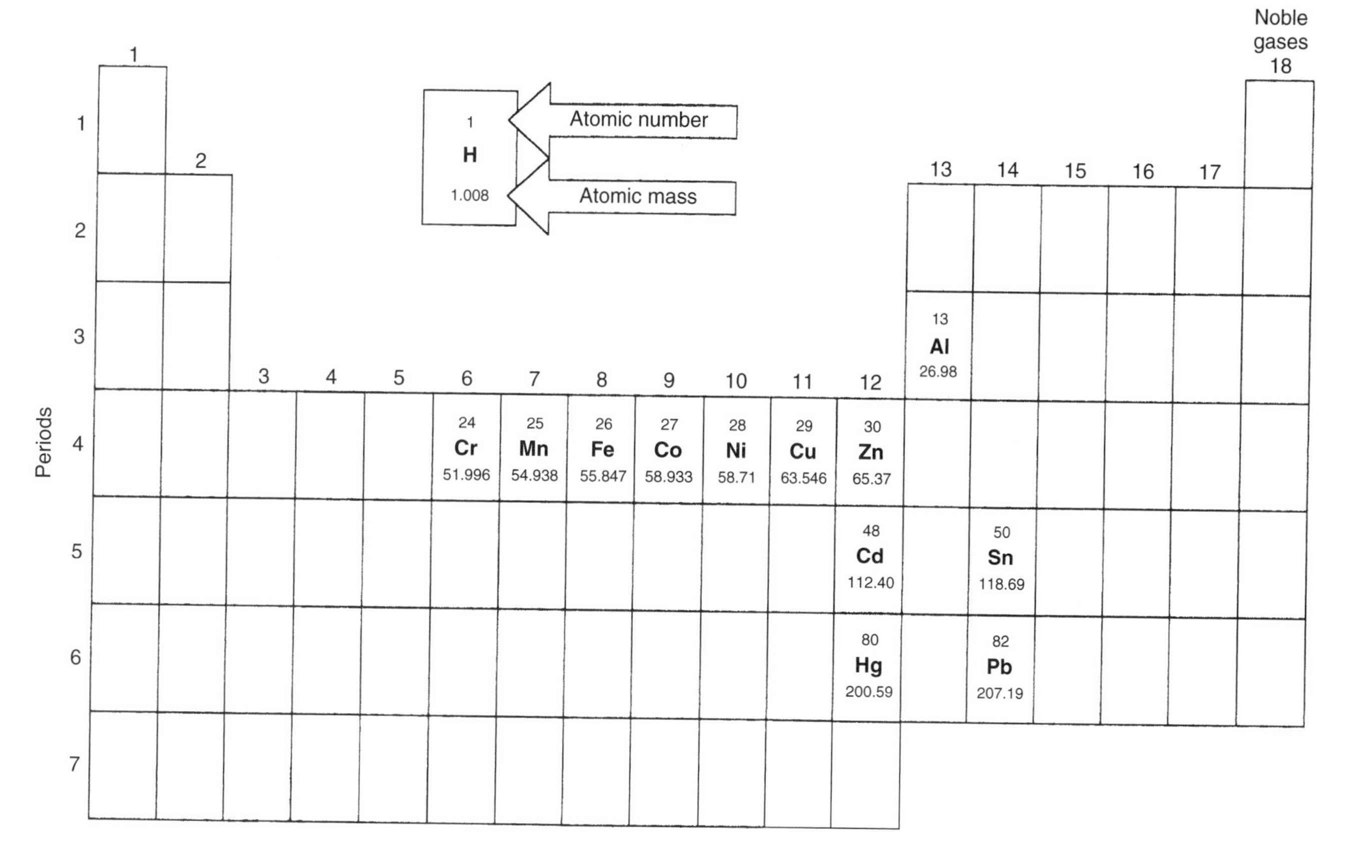

Figure 2.8 Metals of common environmental concern as found in the Periodic Table.

The following paragraphs introduce you to the chemical principles which can govern the transportation of metals in the aquatic environment and give indications as to where high concentrations may be found.

2.4.1 Solubilization

Metals entering the environment are often in an insoluble form in industrial waste, in discarded manufactured products, or as part of naturally occurring mineral deposits. Deposition from the atmosphere is often in the form of insoluble salts. However, the solubility of metals increases with a decrease in pH. Some of the problems of 'acid rain' in causing the death of fish have been attributed to the leaching of toxic metals from the soil, as well as the direct effect of pH on the fish. The use of lead pipes for domestic water supplies is more problematic in areas of soft, acidic water than where the water is hard and slightly alkaline.

Solubilization is often aided by the formation of complexes with organic material. These may be *anthropogenic* (e.g. complexing agents in soap powders) but may also occur naturally. Humic and fulvic acids produced by the decay of organic material can help solubilize metals.

2.4.2 Deposition in Sediments

This can occur when there is an increase in pH. The pH at which this occurs may vary from metal to metal, although under sufficiently alkaline conditions all transition metals will precipitate. Deposition of relatively high concentration metals may result in traces of other metal ions also being deposited. This is known as *co-precipitation*. Metal ions may also interact with sediments by a number of mechanisms, including the following:

- adsorption
- ion exchange (clay minerals are natural ion exchangers)
- complex formation within the sediment

A change in the oxidizing or reducing nature of the water (i.e. the redox potential) may lead either to solubilization or deposition of metal ions. Most transition metal ions can exist in a number of different oxidation states in solution (e.g. iron can exist as Fe^{2+} and Fe^{3+}). Iron in solution under slightly acidic conditions is predominantly Fe^{2+}. Under alkaline oxidizing conditions, the iron is oxidized and precipitates as $Fe(OH)_3$. Under reducing conditions, all sulfur-containing ions (e.g. $SO_4{}^{2-}$) are reduced to S^{2-}, and this may lead to the deposition of metals such as lead and cadmium as their insoluble sulfides.

2.4.3 Uptake by Organisms

From the above considerations, an obvious route into the food chain is from sediments via filter feeders. Many metals are retained in the organism as a simple ion. Others, particularly cadmium and mercury, can be converted into covalent

Table 2.2 Concentrations of trace elements in individual organs of shellfish

Sample	Percentage of whole animal	Concentration (mg kg^{-1})	
		Pb	Cd
Scallop			
Gills	10	52	<20
Muscle	24	<5	<20
Fatty tissue	17	8	2000
Intestine	1	28	<20
Kidney	1	137	<20
Gonads	20	78	<20
Sediment	–	<5	<20
Sea water (mg l^{-1})	–	3	0.11

organometallic compounds. These will behave in a similar fashion to the covalent organic compounds described previously and will preferentially accumulate in fatty tissues. The distribution of the metal within an organism is thus very dependent on the individual metal and its detailed chemistry. Compare the distribution of lead and cadmium in shellfish in Table 2.2.

DQ 2.3

One metal which is of current environmental concern cannot be described either as a transition metal or as a 'heavy' metal. What is this metal?

Answer

Aluminium, which is found in great abundance in the aluminosilicate structures of clays, but is usually fixed in this insoluble form. When the acidity increases sufficiently, this solubilizes the aluminium.

SAQ 2.3

Compare the routes by which high-molecular-mass organic compounds and toxic metals may disperse and reconcentrate in the environment and in organisms.

2.5 What is a Safe Level?

We have now discussed many of the concepts needed to determine the movement of pollutants in the environment and, if degradation of the compound is slow, how reconcentration may occur.

Interpretation of the analytical data needs to be based on the relationship between the analytical concentration and the effect on organisms. This correlation may not be as easy to determine as first may be thought.

Toxicological testing has been performed on many (but by no means all) compounds which produce major environmental problems. The testing is generally under short-term, high exposure ('acute' exposure) conditions. This may take the form of determining the dose or concentration likely to cause death to a percentage of test organisms. The 'LD_{50}' test, for example, determines the lethal dose required for the death of 50% of the sample organisms. This testing is, however, not generally relevant to environment problems, where it is much more likely that the exposure is over a long term in small doses or low concentrations ('chronic' exposure). The effect may be non-lethal, such as a reduction in the rate of growth or an increase in the proportion of mutations in the offspring, but over several generations still leads to a decrease in population of the species. Monitoring of chronic effects may not be easy outside of the laboratory, and may be complicated by the presence of other pollutants, or other uncontrollable effects (e.g. climate). One of the reasons why the environmental problems of p,p'-DDT are often discussed is that its initial release in the early 1940s was into an environment largely free from similar pollutants. Possible effects could be readily correlated with analytical concentrations. This is not as easy nowadays as any compound under investigation will invariably be present in organisms as part of a 'cocktail' with other compounds.

This leads us to the next problem that the effect of two or more pollutants together may be greater (*synergism*) or less (*antagonism*) than that predicted from the two compounds individually. For instance, the effect of sulfur dioxide and dust particles in some forms of smog is much greater than the separate effects of the two components. The toxicity of ammonia in water decreases with a decrease in pH (i.e. with an increase in the hydrogen ion concentration). The ammonium ion, which is the predominant species under acidic conditions, is less toxic than the non-protonated molecule predominating under alkaline conditions.

The consequence of this for the interpretation of analytical data is that information on the concentration of secondary components is often as important as the major analysis. This complicates the analytical task significantly.

2.6 Sampling and Sample Variability

2.6.1 Representative Samples

Before we discuss chemical analysis, we need to consider what could be considered as being a representative sample. It is sometimes not appreciated how variable the environment and its contamination may be. No two living organisms will have had exactly the same exposure to a pollutant and this will give different concentrations in the body of each organism. Effluent concentrations may vary

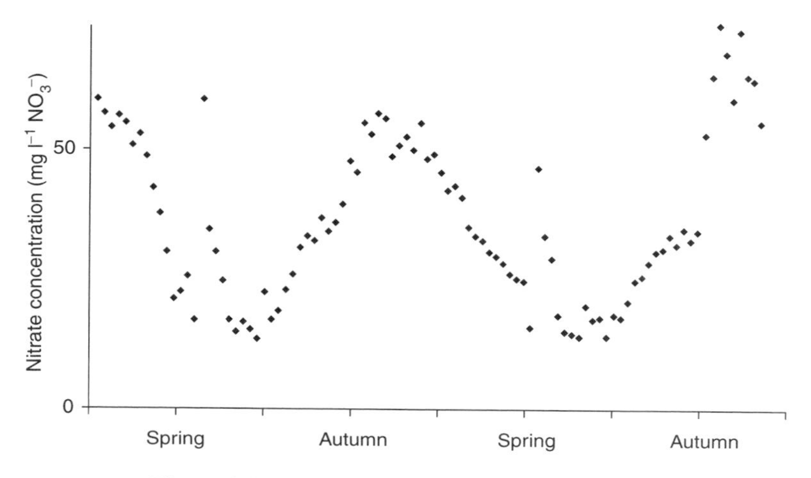

Figure 2.9 Typical variation of nitrate in a river.

if a factory does not operate at night or over the weekend or if the process producing the effluent is not continuous. Concentrations in soil can be different even in adjacent samples. With water or atmospheric samples the concentrations may change hour by hour, day by day or with the seasons. If you have a look at Figure 2.9, which shows a typical variation of nitrate concentrations at a single location in a river, you will be able to see a cyclical variation over the year. Even some consecutive sampling points are significantly different, thus showing a large short-term variation. Different analytical results would be found a few kilometres downstream due to transfer of components into and out of the river and the chemical and biological reactions taking place within the river. Any comprehensive sampling strategy would involve taking a number of samples at different times and from different locations to take into account this variability. The strategy will be discussed in each of the following chapters for specific analytes. Analytical results obtained from single samples may have very little meaning.

2.6.2 Sample Storage

Once the samples are taken they must be kept in such a manner that the concentration of the species to be analysed is unchanged during transportation and storage. Problems may occur if the analyte is volatile, degradable, reactive towards other components in the sample or can deposit on the container walls. Leaching of compounds from the container walls (metal ions from glass containers and organic compounds from plastic containers) may introduce contaminants into the sample. Storage procedures will be different for each sample type and compound being analysed. These problems will be discussed in each of the following chapters before the relevant laboratory analytical procedures are described.

The importance of correct sampling and sample storage cannot be overestimated as no matter how sophisticated the available analytical equipment may be, it can only analyse the sample that is brought into the laboratory. The phrase often used when inaccurate data are sent for computer analysis, i.e. 'rubbish in......rubbish out', is just as applicable to chemical analysis!

2.6.3 Critical Paths and Critical Groups

By using the arguments presented in the previous sections, you should now be able to predict routes by which a particular compound may be transported through the environment. We could start to predict which types of organisms would be most affected. This is a necessary preliminary step for any new monitoring programme in order to maintain the sampling within practicable limits. Even so, the analytical task could still be enormous. When the programme has become established, the use of **critical paths** and **critical groups** can reduce the task. The critical path is the route by which the greatest concentration of the pollutant occurs, and the critical group is the group of organisms (or people!) most at risk at the end of the critical path. If the concentration of the compound in samples taken from the critical group is within the permitted range, then it follows that the concentrations will be no higher in other groups. Monitoring can be largely directed towards the assumed critical path and group but more widespread monitoring should continue – you may be wrong in your choice of path, or the conditions for which you deduced the path may change. The continuing programme should check these assumptions.

Of course, the above is just one example of the many types of analysis which you may have to undertake. Have a look back at the more complete range in Section 1.4 and in the answer to SAQ 1.2 for an appreciation of the diversity of environmental samples.

SAQ 2.4

Consider the discharge of aqueous waste into a semi-enclosed sea area in which there is a thriving in-shore fishing industry. If the waste consists of low-concentration transition and actinide metal salts, what would be the likely critical path and critical group of people?

2.7 General Approach to Analysis

We have seen how many ions and compounds can build up in concentration in organisms even when the background concentrations are in the $\mu g\ l^{-1}$ range. In some instances where the compound is highly toxic, resistant to biodegradation, and bioaccumulates very readily, concern is expressed even when the concentrations approach the limit of experimental detection. This is the case with the dioxin and PCB groups of compounds which are routinely monitored at ng l^{-1} concentrations.

At the other end of the concentration scale, monitoring is often required in water for components which may be present in tens or hundreds of mg l^{-1}. In these cases, the analysis may not necessarily be specific to individual ions or compounds as the measurements are often concerned with the bulk properties of the water (e.g. acidity and water hardness). These are often known as 'water quality' parameters.

DQ 2.4

From your knowledge of analytical techniques, list briefly the types of method which may find use in environmental analysis for organic compounds and metals.

Answer

The broadest categories which you may have listed are probably:

(*a*) *classical methods of analysis, i.e. volumetric methods and gravimetric methods;*

(*b*) *instrumental methods.*

You may then have subdivided the instrumental methods, but we will start with these broad divisions.

Volumetric analyses (titrations) are rapid, accurate, use simple and inexpensive apparatus, and can be used for direct measurements of the bulk properties. Water hardness, for instance, can be measured by a single titration, regardless of the nature of the ions producing the effect. They are, however, of limited use for concentrations below the mg l^{-1} concentrations, and (although automation is possible) can be labour-intensive. Gravimetric techniques can be of extreme accuracy, but very prone to interference from other species. A high degree of skill is necessary for accurate analyses. These tend to be slow techniques due to the time taken for precipitation, filtration and drying. In the few instances where they are used, gravimetric methods are used as reference methods to check the accuracy of other techniques.

Instrumental methods are usually more suited to low concentrations. The linear operating range (i.e. the range in which the reading is directly proportional to the concentration) of instrumentation is generally at the mg l^{-1} level, often corresponding very closely to environmental concentrations. The analysis of the sample is generally rapid, and can easily be automated. You should, however, be aware that sample preparation time and instrument calibration, if not themselves automated, can often be more time-consuming. Accuracy is lower than for the classical techniques, although sufficient for most applications. The majority of the instrumental methods we will be discussing fit into one of the following categories:

- chromatographic methods
- spectrometric methods
- electrochemical methods

As already mentioned, the methods may be sufficiently sensitive for many analyses, often with little sample preparation. Preconcentration of the sample may be used to decrease the lower detection limits of the techniques. In addition, a pre-analytical separation stage may be included to remove interfering components. We can then construct a typical analytical scheme which will cover many of the methods discussed in later sections, as follows:

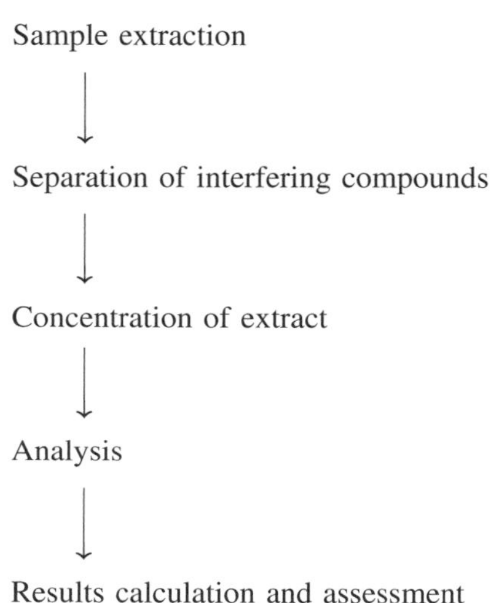

You will be able to see that the analytical stage is just one in what can sometimes be a long series of steps. In the following chapters, each step is described for the different sample types of interest. You will find that there have recently been several advances in terms of sample preparation and in the sensitivity of the analytical stage. These new techniques tend to be faster and more capable of automation than the well established existing techniques and are aimed at high-throughput laboratories. Where new techniques are described, they are compared and contrasted with the existing methods.

2.8 The Choice of Laboratory or Field Analysis

The vast majority of analyses on water or solids are performed on samples taken to a laboratory. There may, however, be circumstances where field analyses are preferable. Atmospheric analysis is often at the point of sampling.

DQ 2.5

What do you think are the relative merits of laboratory and field analyses?

Answer

In the laboratory, the analyses are performed under optimum conditions which will lead to maximum accuracy. Since such analyses are performed in one location, perhaps with a single apparatus, precision will also be maximized. Samples do, however, need to be taken and transported back to the laboratory. There is a time delay in producing the results. Errors may occur from changes which may occur to the sample during storage either by reaction, by loss of sample or by contamination. Laboratories are expensive to build and operate.

Field analysis will produce instantaneous results, although the analytical conditions under which they are measured may be far from optimum, even on a sunny and dry day! Analytical accuracy and precision will be expected to be lower than for a laboratory analysis, but errors due to sample storage will be removed.

There is the possibility, with suitable equipment, of continuous monitoring in the field. This is obviously not possible with laboratory analyses.

An advantage of field sampling which you would probably not have considered is that it may be possible to analyse for species *in situ* which are so reactive that they would not survive transportation to the laboratory. This is particularly the case with reactive atmospheric components.

Field analysis could use the following equipment:

(i) *Portable monitors for specific ions or compounds.* Simple monitors (mg/l concentration range) have been available for water samples for many years and have found large-scale use with organizations which need rapid and simple tests for water quality. Newer types of monitors can determine pollutants at μg/l concentrations and are often used for screening samples to minimize the number of expensive laboratory analyses. Portable gas monitors are standard for health monitoring and for site analysis. Instruments are available to detect specific pollutants in contaminated and reclaimed land.

(ii) *More complex instruments which can be left at secure locations or used in mobile laboratories.* These have long been used in air analysis where they can be part of networks for monitoring air quality. Mobile laboratories find use in urban atmosphere investigations and for contaminated and reclaimed land. Continuous monitoring is sometimes undertaken on major rivers, e.g. the Thames and the Rhine. You could also include in this category ship-board laboratories used for marine investigations.

(iii) *On-line monitors for discharge pipes or flue gases*. These may be used to warn of high concentrations in flue gases or aqueous discharges for compliance with the relevant legislation.

You should not underestimate the degree of sophistication needed for instruments which must operate automatically in the field. At the design stage, you would have to consider minimization of the use of consumables, ensuring that the sampling system never becomes blocked and that the measurement device (often spectrometric or electrochemical) can be kept continuously clean. The instrument should be self-calibrating and the results should be automatically logged or transmitted to a remote location.

Although the main route for chemical analysis is by laboratory analysis, field analysis is playing an important and increasing role, particularly for screening. Examples are described in subsequent chapters following discussions of the laboratory methods.

2.9 Quality Assurance

We have already learnt that pollutants can have concentrations in the environment of $\leq\mu g/l$ and these concentrations can vary widely. Samples for analysis can be water, the atmosphere, solids or living organisms. Whatever the sample or concentration, it is important to be confident in the analytical method and the result produced.

Let us now consider think what the term 'confidence' could mean.

(i) The method used should have been validated prior to the analytical investigation. i.e. thoroughly tested to show that the method gives accurate results for the type of sample being analysed. Ideally, the method itself would include procedures to confirm its reliability for each fresh batch of samples.

(ii) There is some indication of the error inherent in the method.

DQ 2.6

Why do you have to confirm reliability for **each fresh batch** of samples?

Answer

Potential interferences may change from sample to sample. There may also have been changes in the reagents used or procedures within the laboratory which may affect the result.

If we are concerned about the accuracy of a result it is obvious that concern should extend all the way from sampling to the publication of the final analytical result.

DQ 2.7

What areas would you consider to be important in producing an accurate analysis?

Answer

- *The sampling procedure should produce a representative sample.*
- *The sample should not become contaminated or alter chemically during storage.*
- *There should be no contamination of the sample within the laboratory or during the analysis.*
- *Any losses in extraction, separation and concentration procedures should be minimized.*
- *There should be no interference in the final analysis from other components in the sample.*
- *Results should be correctly calculated and archived for future reference.*

Most of these concerns would be applicable to any area of analytical chemistry but the potential contamination of the sample during sampling, sample storage or in the analytical determination is of particular importance in environmental analysis. Many of the compounds are universal contaminants and so will be found in the materials used for sample containers, the apparatus used, solvents and even in the laboratory atmosphere. It may be surprising for you to realize that reliable data for the concentrations of trace metals in sea water have only been available for the last two decades. The values that are now accepted can be an order of magnitude lower than the previous 'best' figures. The earlier values were very largely due to the metal ions picked up during the analytical procedure.

A second major area of concern in environmental analysis is that of interfering compounds. When working at trace or ultra-trace levels, it is easy for there to be components in the sample which remain unseparated from the analyte even after extensive pretreatment. This would lead to an increase in the analytical result above the true value. From the nature of many environmental samples it can often be very difficult to predict what these potential interferences would be. The samples could include many unexpected components.

You may have already come across the terms Quality Assurance and Quality Control. Their precise definition varies between organizations and countries, although the following would be generally acceptable:

- Quality Assurance – the overall methodology needed to minimize the potential errors.
- Quality Control – the measures used to ensure the validity of individual results.

Table 2.3 Some examples of quality assurance procedures

- Sampling and sample storage procedures which ensure that the sample is truly representative and that it reaches the laboratory unchanged.
- Sampling and analysis in duplicate.
- Specifications within the analytical scheme for reagent purity and apparatus cleanliness.
- Repeated checks on the instrument performance or chromatographic resolution.
- Traceability in any standards used. This means that the stated concentrations in any standard used must be traceable back to primary international standards.
- Inclusion in each analytical batch of additional samples of known composition. These will confirm the reliability of the method and could include the following:

  Blank samples – samples made up as close as possible in composition to the unknown, excluding the compound being determined. These are introduced before stages in the analysis when contamination is likely. A positive determination of the analyte in the blank would indicate contamination.

  'Spiked' samples – these are samples to which a known quantity of the compound being determined has been added. A valid analysis of the spiked and unspiked sample will be able to determine accurately the quantity added.

  Reference samples – these are materials which are similar in type to the unknown sample and have an accurately determined composition.

Whenever you are assessing an analytical scheme, look out for quality assurance steps in the analytical procedures. Some examples of these are given in Table 2.3.

2.9.1 Finding a Suitable Method

For routine monitoring you would probably find standard methods available from various national or international organizations. These methods usually detail not only the experimental procedures but also their range of applicability (concentration range and sample type), limits of detection and expected errors. The organizations producing such methods include the following:

- The American Society for Testing and Materials (USA)
- The British Standards Institute (UK)
- The Environment Agency (UK)
- The Environmental Protection Agency (USA)
- The International Standards Organization

For less routine work, you may have to search the literature for investigations similar to your own. The techniques used may need some modification. For example, you may discover a technique which had been investigated and validated for sea water which you need for monitoring fresh water. You should revalidate the methods for your own investigation and sample type before starting the new analytical programme.

2.9.2 Laboratory Standards

You have not quite finished in your task to ensure production of a reliable analytical result. Other factors can include the following:

- General cleanliness of the laboratory
- Contamination of the laboratory equipment and atmosphere from previous analyses
- Training of the laboratory staff
- Frequency of instrument maintenance and calibration

Many countries have protocols or certification programmes which attempt to ensure that these problems are minimized. Examples include the following:

- The National Accreditation Management Service (NAMAS) in the UK. This accredits laboratories for specific analytical procedures.
- Good Laboratory Practice (GLP) assesses laboratories themselves to work within a defined scientific area. This scheme was devised by the Organization of Economic Co-operation and Development (OECD).

It may be thought that this section on quality seems all rather obvious, i.e. just repeating the good practice that a competent and conscientious analyst would in any case be doing? The problems when working at low concentrations may be easily underestimated. During the validation of new nationally or internationally recognized procedures, there are often inter-laboratory tests. It is not unknown for well-established and reputable laboratories to produce results which are well outside the expected error range. Perhaps the permitted purity concentrations of the laboratory reagents were not low enough or a critical step in the analysis not carefully enough defined. While it can be relatively easy to produce a numerical result from an analytical procedure, it sometimes is very difficult and requires considerable effort to produce an **accurate** numerical result.

SAQ 2.5

A number of samples to be analysed for traces of a common solvent are taken from a river flowing through a highly polluted area. The samples are transported to the laboratory for analysis by gas chromatography. Which steps of the procedure would need to be monitored in order to ensure that the sample was not contaminated? What quality control procedure could you introduce to ensure reliability of the analytical result?

Summary

Pollutants travel through the environment by routes which can be predicted from their chemical and physical properties. High-molecular-mass neutral organic compounds and many metals are of considerable concern. Such species are capable of reconcentrating in certain areas and within organisms and it is in these areas where they have their greatest effect. An understanding of such routes is needed for the correct choice of sampling positions for the subsequent analytical determinations. Analysis is normally carried out in a laboratory, although field analyses can sometimes be found useful. An introduction is given in this chapter to the available techniques and the quality assurance necessary to produce reliable data.

Chapter 3

Water Analysis – Major Constituents

Learning Objectives

- To list the major constituents of environmental waters, their concentrations and to describe how the concentrations may change during passage through the environment.
- To appreciate the importance of correct methods of sampling and sample storage.
- To be able to describe methods for the measurement of water quality.
- To determine the most suitable analytical techniques for the analysis of the major constituents of water.

3.1 Introduction

Water is vital for life. Not only do we need water to drink, to grow food and to wash, but it is also important for many of the pleasant recreational aspects of life.

DQ 3.1

List the uses which we can make of water.

Answer

This should include the following, although you may have thought of some extra ones of your own:

- *Domestic water supply*
- *Industrial water supply*
- *Effluent and waste disposal*
- *Fishing*
- *Irrigation*
- *Navigation*
- *Power production*
- *Recreation, e.g. sailing and swimming*

Each different use has its own requirements over the composition and purity of the water and each body of water to be used will need to be analysed on a regular basis to confirm its suitability. The types of analysis could vary from simple field testing for a single analyte to laboratory-based, multi-component instrumental analysis. Water is found naturally in many different forms. In the liquid state it is found in rivers, lakes and groundwater (water held in rock formations), and also as sea water and rain. As a solid, it is found as ice and snow. Water in the vapour state is found in the atmosphere. You will certainly be familiar with the fact that sea water contains large quantities of dissolved material in the form of inorganic salts but it may come as a surprise that nowhere in the environment can you consider water to be chemically pure. Even the purest snow contains components other than water.

DQ 3.2

Write down some of the constituents which you consider might be found in natural river water.

Answer

- *Ions derived from commonly occurring inorganic salts, e.g. sodium, calcium, chloride and sulfate ions.*
- *Smaller quantities of ions (e.g. transition metal ions) derived from less common inorganic salts, perhaps derived from leaching of mineral deposits.*
- *Insoluble solid material, either from decaying plant material, or inorganic particles from sediment and rock weathering.*
- *Soluble or colloidal compounds derived from the decomposition of plant material.*
- *Dissolved gases.*

You will probably have written down most of these. The category which many forget to include is the 'dissolved gases'. This, of course, includes oxygen which is so vital in supporting aquatic life. Dissolved gases occur through contact with the atmosphere and through respiration and photosynthesis. A fast flowing turbulent river will usually be saturated in atmospheric gases. Respiration of aquatic animals releases energy from foodstuffs, consuming oxygen and producing carbon dioxide, as follows:

$$\underset{\text{glucose}}{C_6H_{12}O_6} + 6O_2 \longrightarrow 6CO_2 + 6H_2O + \text{energy} \quad (3.1)$$

Photosynthesis by plants reverses this process, producing organic compounds and oxygen from carbon dioxide by using sunlight as an energy source:

$$6CO_2 + 6H_2O + h\nu \longrightarrow C_6H_{12}O_6 + 6O_2 \quad (3.2)$$

Oxygen levels in water are depleted by slow oxidation of organic and, in some cases, inorganic material. The presence of large quantities of oxidizable organic material (e.g. from sewage effluents) is often the most serious form of pollution in watercourses.

Ions commonly found in the mg l^{-1} concentration range are shown in Table 3.1. Others (e.g. fluoride ions) may occur depending on the mineral deposits in the locality.

Your list should also have included the compounds derived from decomposition of plant material. Did you include inorganic as well as organic products, as shown in Figure 1.1? Don't forget ammonia. This can occur in water in the 0–2 mg l^{-1} range. Concentrations never usually increase to greater than these values as ammonia is rapidly oxidized to nitrate. It has significant toxicity to fish, particularly when it is present as the neutral molecule, rather than when protonated to form the ammonium ion.

Now look at Figure 3.1, which shows typical comparative analyses for rain water, river water and sea water. You will find similar ions in all three, with the only difference being the concentration range. Sea water contains the common ions at the g l^{-1} level, whereas for river and rain water the values are at the mg l^{-1} level. All are easily measurable with modern instrumentation.

The situation would be a little different if we tabulated the less common species. The range of ions (particularly metal ions) would be limited in river water by the chemical composition of the rocks over which it was flowing. On the other hand, sea water contains trace quantities of virtually every element, with the highest concentrations being found close to the surface and in coastal areas. This is a very complicated analytical matrix indeed.

Table 3.1 Ions found in mg l^{-1} concentrations in natural waters

Concentration range (mg l^{-1})	Cations	Anions
0–100	Ca^{2+}, Na^+	Cl^-, SO_4^{2-}, HCO_3^-
0–25	Mg^{2+}, K^+	NO_3^-
0–1	Fe^{2+}, Mn^{2+}, Zn^{2+}	PO_4^{3-}
0–0.1	Other metal ions	NO_2^-

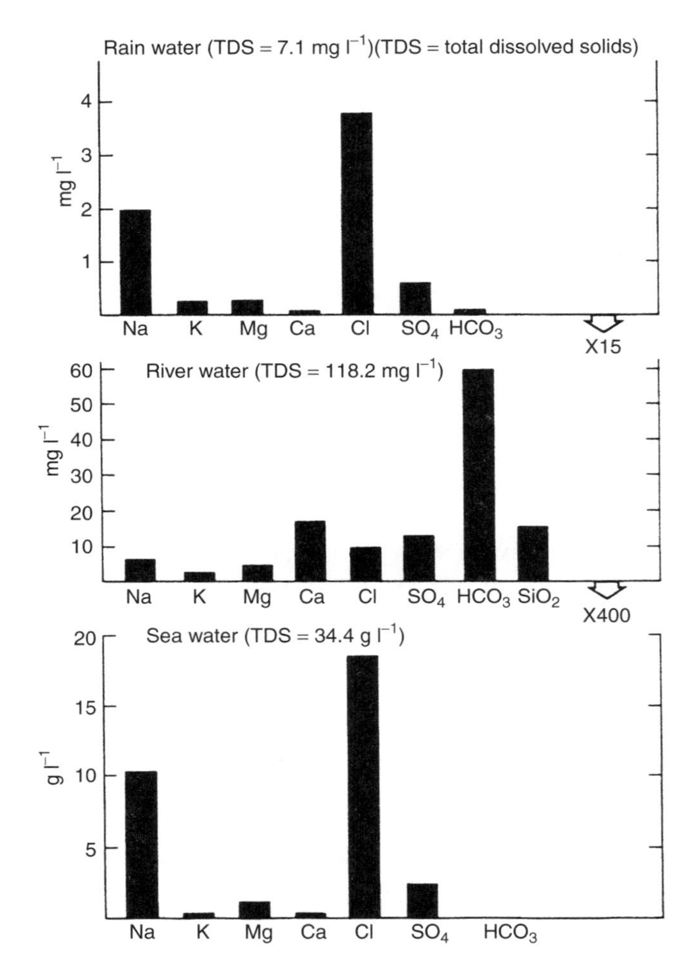

Figure 3.1 Typical comparative analyses for rain water, river water and sea water; note the different scales for each histogram. Reprinted with permission from Gibbs, R.J., *Science*, **170**, 1088–1090 (1970). Copyright (1970) American Association for the Advancement of Science.

Have you noticed that, although the absolute concentrations within rain water and sea water are very different, the relative concentrations are often very similar, thus giving us a clue as to the origin of these ions?

DQ 3.3

Estimate and comment on the concentration ratios of the following in the rain, river and sea water concentrations:

Calcium/Chloride and Chloride/Sulfate

Answer

Rain, especially when falling close to the sea, has sea water as a major constituent and can often be regarded as diluted sea water.

A detailed comparison of the concentration of ions from a large number of rivers, compared with the concentrations in sea water (which appears depleted in a number of elements, including calcium), is one of the methods of studying the complexities of marine chemistry. Unfortunately, further discussion of this is outside the scope of this present book.

Water authorities often feel it necessary to analyse a river at many locations along its course. This is because the composition of water is never static. It changes by interaction with the atmosphere and crust, and by chemical and biological processes occurring within the water. This does not even include the possibility of extra material being added in the form of pollution. Let us consider a river flowing from its source to the sea. Even at its source, water will contain dissolved salts from the passage of water through the earth to form the river. Some of the natural processes which will affect the constituents are listed below and are also illustrated in Figure 3.2.

(i) *Weathering of rocks.*
This will produce an increase in inorganic salt content. The composition may also be affected by interaction with material on the river bed. Clays, often found on river beds, are natural ion exchangers.

(ii) *Sedimentation of suspended material.*
As the river progresses downstream it will generally become less turbulent and so less capable of supporting suspended material.

(iii) *Effect of aquatic life.*
Consumption and production of oxygen and carbon dioxide by plants has already been mentioned. Living plants will also absorb nutrients (including nitrate and phosphate) necessary for growth.
The death and decay of organisms will release ions and also produce suspended material. This will slowly decompose into simpler chemical compounds. If the process proceeded to completion in the presence of oxygen, the final products would be carbon dioxide and water. At the same time, the oxygen concentration would fall. If the oxygen concentration was already low, then the final products would include ammonia and methane.
Dense beds of vegetation can also very effectively filter out suspended solids.

(iv) *Aeration.*
The generation of oxygen by plants is not the only method by which the gas enters water. There is continuous transfer of gases between the atmosphere and water. The oxygen can replenish the oxygen removed by oxidation of organic material.

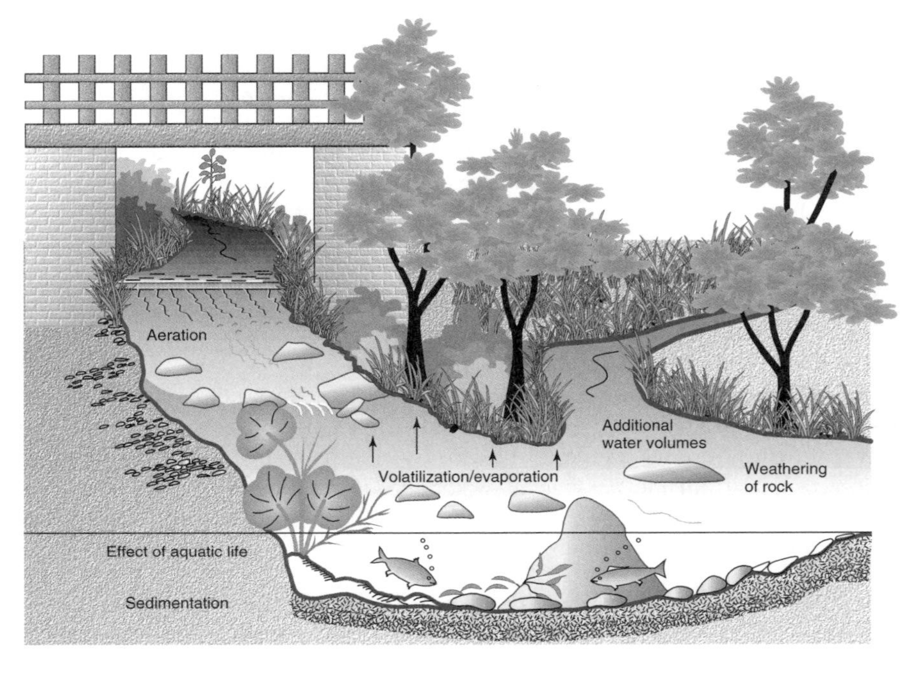

Figure 3.2 Natural processes affecting river composition.

(v) *Volatilization and evaporation.*
Low-relative-molecular-mass organic compounds tend to have a high vapour pressure and will be readily lost from water. A significant percentage of the water itself in the river can be lost through evaporation (the rate depending on the ambient temperature) and this will have the effect of increasing the concentration of all dissolved material in the river.

(vi) *Additional water volumes.*
Any water entering from tributaries or directly from overland flow will alter the analytical concentrations and may bring new constituents to the river.

Similar considerations should allow you to understand the composition of waters in other areas in the environment.

DQ 3.4

Groundwater is sub-surface water in soils and geological formations where the ground has become saturated with water. If held in permeable rock the water can be extracted for use.

Keeping in mind the passage of water from the surface, how would you expect the composition of groundwater to be different from surface water?

Answer

The groundwater could be more concentrated in salts leached from mineral deposits. During passage through the earth, the water will have been in contact with degradable organic material. This can lower the oxygen content of the water.

Even if you disregard the introduction of new compounds by pollution, any environmental water will contain a large number of components. In fact, if you start considering components which may be found at trace levels (less than mg l^{-1}) the task would be almost impossible as new components are constantly being identified in natural waters. Thankfully, it is very rare that all of the components would need to be analysed.

This present chapter includes methods for the analysis of major components of water which may be routinely undertaken by water authorities. Even so, it would be unusual for all of the methods to be used on one sample. Water authorities or others undertaking the analyses will in general have a reasonable idea of what species to expect in the water. Unless there is a specific reason for more complete analysis, the analytical scheme will usually be restricted to components which are likely to cause environmental problems or exceed prescribed limits. Remember that consideration of the analytical process has to start with sampling and sample storage (see Section 2.6 above).

SAQ 3.1

Using the list you produced of likely chemical species in a river, decide which would be likely to increase or decrease downstream from a sampling point close to the source.

SAQ 3.2

Lakes may have a complicated chemical structure. At some times of the year, some lakes can be well-mixed and so will be chemically homogeneous. At other times, there is little movement of water and the lake becomes stratified with the top surface of less dense water heated by the sun, a central layer and a lower more dense layer of colder water. There is little transfer between the layers. What differences in composition in the water would you expect between the top and bottom layers?

3.2 Sampling

Let us now consider developing a sampling programme for a river.

The sample or samples (often only 250 or 500 ml each) must be representative of the whole body of water requiring analysis. The sample must also be kept in such a manner that the concentration of the species to be analysed is unchanged during transportation and storage.

(i) Before starting, decide on what analyses are required. The analytical techniques to be used will affect the sample size taken, the type of sample bottle and also the method of storage. It will be too late to alter these by the time you get back to the laboratory. You should also confirm that laboratory time is available for analysis of the samples. Sample preservation times should be kept to a minimum (hours to days, depending on the analysis). The maximum holding times may have already been defined in quality assurance schemes (see Section 2.9 above).

(ii) Decide on a sampling programme. We have already discussed how the composition of natural water is always changing (see Sections 2.6 and 3.1 earlier). Sometimes the variation in composition may be periodic:

Seasonal – the concentration is affected by natural growth processes.
Weekly – a pollutant may only be emitted from a factory during the working week.
Daily – the concentration of some components may be changed due to biological processes needing the presence of sunlight.

You may wish to monitor these regular fluctuations but you may be more concerned with the longer-term variation of concentrations. Your sampling programme, the number of samples, and the timing of the sampling will be affected. If you are interested in long-term variations it may be beneficial to take samples at the same stage of each periodic cycle, whereas for short-term variations you would take several samples each cycle.

DQ 3.5

What regular variations would you expect in concentrations of the following:

- Dissolved oxygen;
- Nitrate?

Answer

Oxygen is produced by photosynthesis in daytime but is consumed by respiration or by oxidation of organic material continuously. There will be a continuous but slow replenishment from the atmosphere. A drop in oxygen concentration during the night would be expected.

Variation of nitrate would be more complex. This is a nutrient which is necessary for growth and so if there were no additional inputs it would decrease in the spring growing season and increase in winter; however, if a farmer put an excessive amount of nitrate-containing fertilizer on a neighbouring field, there would be a sudden increase in any river into which the field drained.

You should recall Figure 2.9 which gives an example of seasonal nitrate concentration changes.

(iii) Decide on the total number of samples you are taking, remembering that each location should be sampled in duplicate. Although it is good practice to start by taking as many samples as you feel necessary for complete monitoring, you do also have to take into consideration the time required for the analyses. It is very common to severely underestimate the time involved in the laboratory analyses.
A further consideration if there is to be any statistical treatment of results is that there are sufficient samples for the treatment to be significant.

(iv) Decide on the location of the sampling and the sampling apparatus. If you are to take samples regularly from one location, the first consideration must be ease of access. Remember that the weather may not always be perfect. Surface water sampling requires little sophistication in sampling apparatus (often directly into a sample bottle or a bucket) but the surface may not be the best location for sampling. It may not provide the most representative sample. There also is the possibility of contamination by surface pollutants. Surface contamination can be largely overcome by inserting the bottle upside down in the water and inverting to fill it from just below the surface. Ideally, however, the river should be sampled further underneath the surface, in its main flow and at similar depths for each sample. A simple sub-surface sampler would be a weighted, stoppered bottle on an attachment line. The stopper is removed at the required depth by a cable. More complex designs, such as the Van Dorn sampler shown in Figure 3.3, are open cylinders with valves at each end and produce less disturbance to the river on sampling. The sampler is sealed by using a weight (messenger) dropped down the attachment line to activate the valve mechanism.
If you are monitoring the effect of a discharge into a river, samples should be taken far enough downstream for the discharge to be completely mixed (Figure 3.4). Samples taken further upstream would be unrepresentative as the analysis would depend on how much the discharge had mixed with the river.

(v) Decide on the sample volume to be taken to the laboratory and the sample-storage containers. The latter are usually made of glass or polyethylene. However, these materials (and those in the container top) are not as inert as you might think. Polyethylene containers may leach organic compounds into the sample, while glass bottles can leach inorganic species (sodium, silica and other components of the glass). How much you fill the container is also important. If you are analysing volatile material or dissolved gases, the container must always be full. For other components, it is beneficial

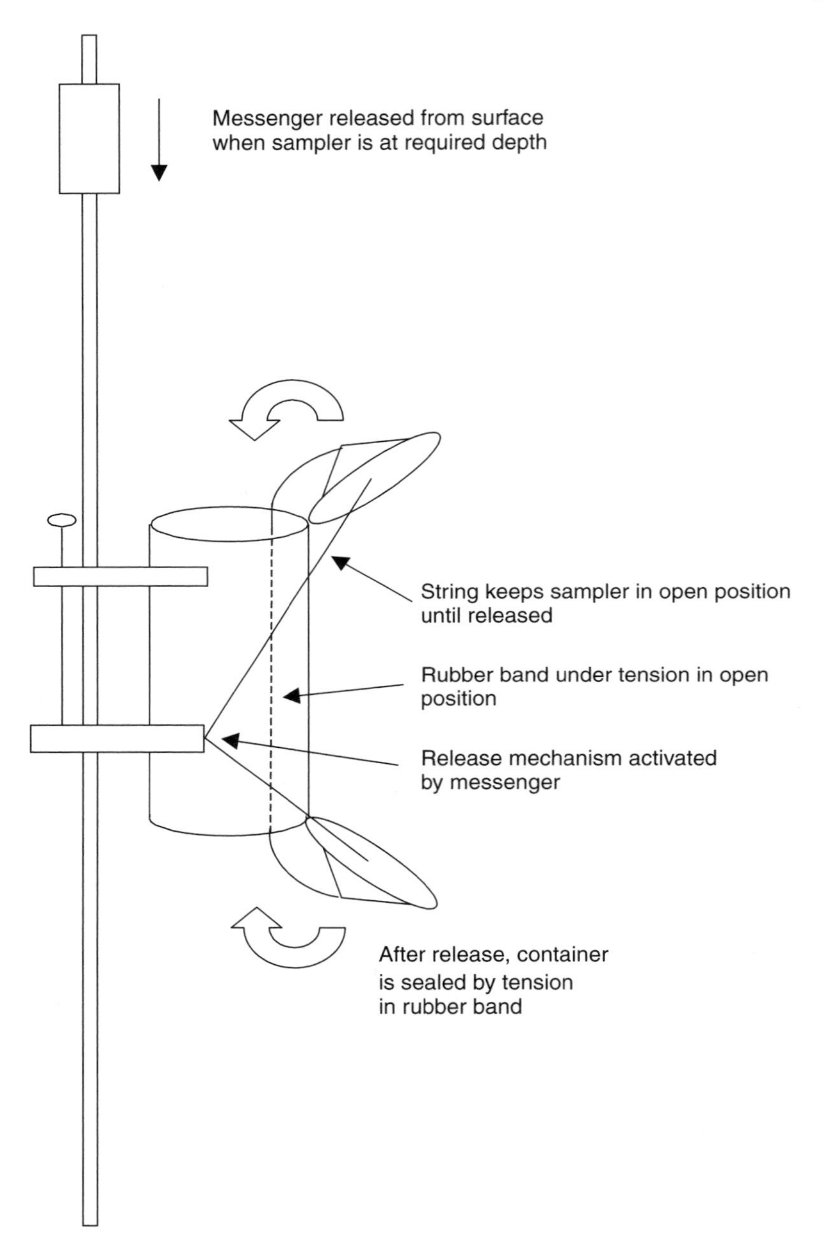

Figure 3.3 Schematic of a Van Dorn water sampler.

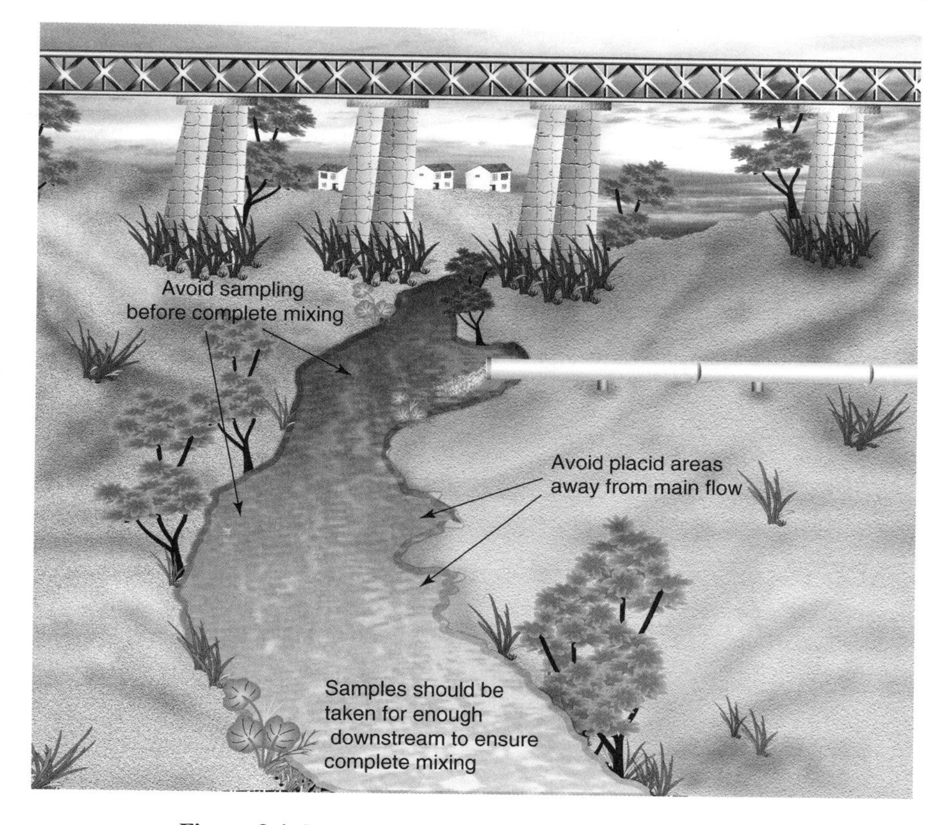

Figure 3.4 Sampling to monitor the affect of a discharge.

not to fill the container completely as the contents can then be more easily mixed before analysis. Try attempting to mix the contents of a completely full container!

At this stage, it would also be worthwhile to check the equipment used at all stages of the sampling procedure to ensure that nothing will introduce contamination – the sampler itself, funnels and any tubing used. If you are sampling from a motorized boat without care, the boat itself could introduce contamination into the water and may disturb the sediment.

(vi) Decide on the method of storage of samples. Standard methods are available for most components to minimize analyte loss. The method varies according to the physical and chemical properties of the species. For example:

Nitrate – store at 4°C to lower biological degradation.

Pesticides – store in the dark to avoid photochemical decomposition.

Metal ions – acidify the sample to prevent adsorption of metal ions on to the sides of the container.

Phenols – add sodium hydroxide to lower the volatility.

For some analyses (e.g. biochemical oxygen demand – see below), no preservation is possible and the analysis should be performed as soon as possible after sampling, keeping the sample cool during transportation to the laboratory.

You should note that this may mean different sample storage conditions and storage containers for each analysis.

After all of these considerations, you can start sampling!

SAQ 3.3

You are about to take samples for the following analyses:

ammonia;
chloroform;
total organic content.

List decisions which have to be made in developing a river sampling protocol. What are the relevant chemical and physical properties which would help you decide on the storage conditions? In addition, suggest storage bottles and precautions to minimize analyte loss.

3.3 Measurement of Water Quality

This section discusses techniques which are usually intended to provide a measurement relating to the overall effect of groups of compounds or ions rather than to measure concentrations of individual components. These were originally conceived as simple and convenient methods to assess waste water but are now in widespread use to monitor long-term changes in environmental waters. You will find that many of the techniques involve titrations or use spectrometry.

3.3.1 Suspended Solids

We can all visualize streams so full of suspended material that the water is opaque, and where no visible life could possibly exist. This represents an extreme case of high solids loading. Any natural water will contain some suspended solids, but often the material is of such a small particle size that it cannot be easily seen. It is only when you look at two samples of water, one of which you consider 'clean' and the other which has been filtered to sub-micron level, that you can see the difference. The filtered water glistens, while the 'clean' water suddenly looks distinctly dirty. Even if the particles are chemically inert, their physical properties could cause problems.

DQ 3.6

Which physical problems do you think may be caused by suspended solids?

Answer

1. *They cut down light transmission through the water and so lower the rate of photosynthesis in plants.*
2. *In less turbulent parts of the river some of the solids may sediment out, thus smothering life on the river bed.*

You may have guessed that the analysis of suspended solids is by filtration and weighing but you may not realize the laboratory skill which is required until you discover that a typical suspended solid loading for a clean looking stream would be only a few mg l^{-1}. Even sewage discharges in the UK have to conform to conditions with a maximum of 30 mg l^{-1} (after 10-fold dilution of the discharge). Typically, a glass fibre filter disc with a 1.6 μm pore size would be used with a Hartley filter funnel (Figure 3.5). The paper is clamped inside the funnel to prevent any part of the sample escaping around the side of the filter.

3.3.2 Dissolved Oxygen and Oxygen Demand

All animal life in a river is dependent on the presence of dissolved oxygen. More subtle requirements for a healthy river also include the presence of oxygen

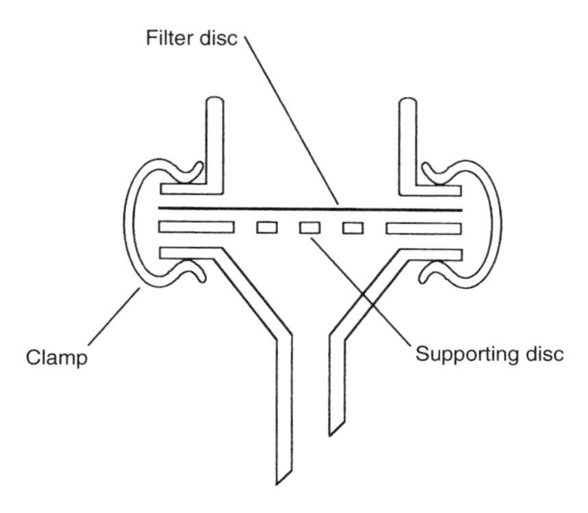

Figure 3.5 Schematic of a Hartley funnel.

for the whole ecosystem. We have already seen how the presence of organic matter can remove oxygen from water by oxidation. Although the process can be written down as a simple chemical reaction, it is, in fact, a microbiological process, known as *aerobic decay*. This converts the major elements present in plant matter (C, H, N, S) into CO_2, H_2O, NO_3^- and SO_4^{2-}, respectively.

It will perhaps come as a surprise that even if no oxygen is present in the water, organic material will still be broken down. Instead of the material being oxidized, it is reduced. The process is once again microbiological and is known as *anaerobic decay*. In this case, the final products are CH_4, NH_3 and H_2S. Consideration of the products of anaerobic decay (in particular, their toxicity, smell and flammability) show that this condition should be avoided at all costs in environmental waters.

The solubility of oxygen in water is low. Saturated water at 25°C and 1 atm pressure contains 8.54 mg l^{-1} oxygen. The sensitivity of fish to low oxygen is very species-dependent. Salmon can only survive under almost saturated conditions, trout to about 1.5 mg l^{-1}, while carp and tench are more resistant, surviving down to about 0.3 mg l^{-1} oxygen. It is easy to deplete the oxygen content if any material is present which would react rapidly with the oxygen. Such material could be organic, as already discussed, but could also be inorganic. Iron in the form of Fe^{2+} can deplete oxygen by oxidation to Fe^{3+}. Natural replenishment by oxygen from the atmosphere can be very slow.

DQ 3.7

Which of the following rivers do you think would take up oxygen most quickly?
Which would be likely to have the highest oxygen demand?

1. A fast-flowing mountain stream.
2. A slowly flowing river in a heavily industrialized area.
3. A slowly flowing river in unspoilt countryside.

Answer

The turbulence caused by the fast flow cascading over rocks in the mountains would ensure that oxygen was taken up rapidly and the water saturated with oxygen. It would be unlikely that the river would contain large quantities of organic matter either from vegetation or from industrial effluent. The oxygen demand would be low.

The slowly flowing river would take up oxygen more slowly as there would be much less turbulence. The heavy industry in the area would be very likely to discharge oxygen-consuming effluent which would increase the oxygen demand of the receiving water.

The river in the countryside would be less likely to contain oxygen-consuming effluent but may still possess a significant oxygen demand from decaying vegetation and also from any material carried downstream into the area.

You should be able to recognize two distinct analyses which could be useful if monitoring environment waters for oxygen:

1. A direct measurement of the oxygen concentration in the sample. This would give an indication of the health of the river at a particular location and at the time of sampling. It would be of less use for assessing the overall health of a river as the oxygen level can vary dramatically with location and with time.
2. A measurement of the amount of material which, given time, could deplete the oxygen level in the river. This is known as the *oxygen demand*, and gives an indication of the possibility of oxygen depletion which will occur if the oxygen is not replenished.
   Such a measurement would be much more suitable for determining the overall health of the river since the oxygen demand is unlikely to change suddenly.

The analytical techniques used for dissolved oxygen measurement can also be used to measure oxygen demand and so these will be discussed first.

3.3.2.1 Dissolved Oxygen

The determination of oxygen can be either by titration (Winkler method) or by use of an electrode sensitive to dissolved oxygen. The results are either expressed as a simple concentration (mg l^{-1}) or as a percentage of full saturation. The concentration of oxygen in saturated water is dependent on the temperature, pressure and salinity of the water and would need either to be established from published tables or determined experimentally. The first problem to overcome is transport of the sample to the laboratory. Without modification to the sample, this would cause sufficient agitation to the water to saturate the sample with oxygen from the air, regardless of its original content.

In the Winkler method, the oxygen is 'fixed' immediately after sampling by reaction with Mn^{2+}, added as manganese (II) sulfate, together with an alkaline iodide/azide mixture:

$$Mn^{2+} + 2OH^- + 1/2O_2 \longrightarrow MnO_2(s) + H_2O \qquad (3.3)$$

The iodide is necessary for the analytical procedure in the laboratory and the azide is present to prevent interference from any nitrite ions which can oxidize the manganese (II) ion. The sample completely fills the bottle to ensure no further oxygen is introduced. After transport to the laboratory, the sample is acidified with sulfuric or phosphoric acid. This produces the following reaction:

$$MnO_2 + 2I^- + 4H^+ \longrightarrow Mn^{2+} + I_2 + 2H_2O \qquad (3.4)$$

The released iodine can then be titrated with sodium thiosulfate using a starch indicator:

$$I_2 + 2S_2O_3^{2-} \longrightarrow S_4O_6^{2-} + 2I^- \quad (3.5)$$

DQ 3.8

What is the equivalence between the original oxygen and the thiosulfate?

Answer

The overall reaction is as follows:

$$2\,S_2\,O_3^{2-} + 2\,H^+ + 1/2\,O_2 \longrightarrow S_4\,O_6^{2-} + H_2\,O \quad (3.6)$$

i.e. four moles of thiosulfate in the final titration is equivalent to one mole of oxygen in the sample.

The electrode method is used for field measurements of dissolved oxygen and can also be employed in the laboratory for determination of the *Biochemical Oxygen Demand* (see below). Several types of systems are available for this purpose, including the Mackereth cell shown in Figure 3.6. In the latter, the current generated by the cell is proportional to the rate of diffusion of oxygen

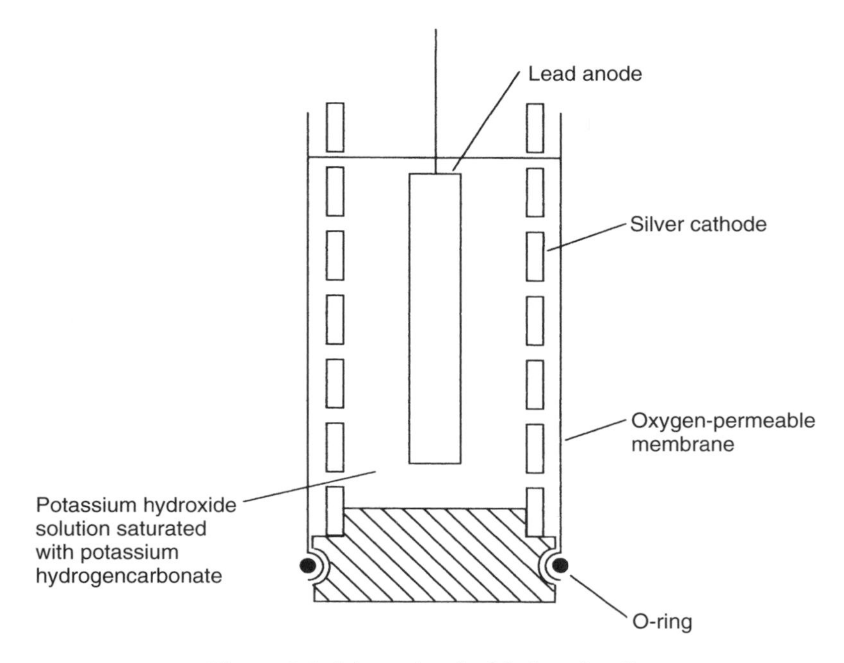

Figure 3.6 Schematic of a Mackereth cell.

through the membrane, which is in turn proportional to the concentration of the oxygen in the sample. The reactions involved are as follows:

at the cathode $$1/2O_2 + H_2O + 2e \longrightarrow 2OH^- \quad (3.7)$$

at the anode $$Pb + 2OH^- \longrightarrow PbO + H_2O + 2e \quad (3.8)$$

Instruments usually read oxygen directly with a scale from 0–100% saturation and are calibrated by setting 100% with fully aerated water and 0% with water with no oxygen content (sodium sulfite is added to the water). This calibration must be made each time that the electrode is used.

3.3.2.2 Oxygen Demand

This can be measured by a number of methods. We will compare these after each of them has been described.

Biochemical Oxygen Demand. The method used to measure the biochemical oxygen demand (BOD) attempts to replicate the oxidation conditions found in the environment. In this, the dissolved oxygen level of a fully aerated water sample is first determined by either of the methods previously described. The measurement is repeated on a sample after it has been left for five days in the dark in a completely filled container and under standard conditions designed to be ideal to promote microbiological activity (20°C, after adjustment of the pH to between 6.5 and 8.5, with the possible addition of salts containing magnesium, calcium, iron (III) and phosphate as nutrients). Care has to be taken with contaminated or treated waters that no compounds are present which would lower the microbial activity, e.g. chlorine. The latter can be removed by the addition of sodium bisulfite.

If the sample is expected to have a high oxygen demand, a dilution should be made with well-aerated water (whose oxygen content is known). Ideally, 30% or more of the oxygen should remain at the end of the analysis. The diluent should include nutrient salts, with distilled water alone not being satisfactory. If, for any reason, the sample is thought to be sterile, a seed sample of sewage may be added.

If there is no dilution of the sample, then we can write the following:

$$\text{BOD} = (\text{initial oxygen concentration} - \text{final oxygen concentration})\text{mg l}^{-1} \quad (3.9)$$

Typical BOD values for unpolluted water are of the order of a few mg l^{-1}. Many seemingly innocuous effluents have a very high oxygen demand, as shown in Figure 3.7. If you remember that the saturated oxygen level in water is of the order of 8 mg l^{-1}, then you will be able to see how the introduction of a small quantity of high-strength effluent can deplete the oxygen in many times its own volume of water.

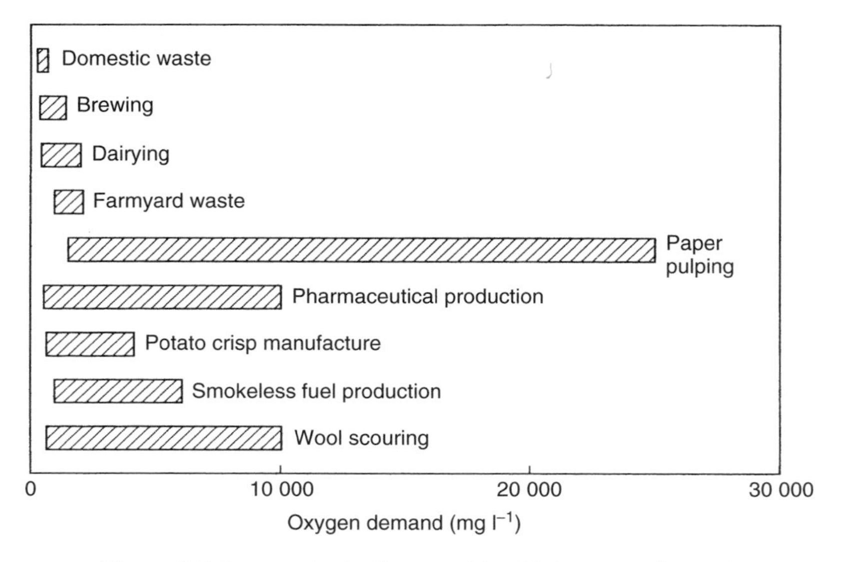

Figure 3.7 Some typical effluents with a high oxygen demand.

Chemical Oxygen Demand. The term 'Chemical Oxygen Demand' (COD) relates to a family of techniques which involve reacting the sample with excess oxidizing agent. After a fixed period, the concentration of unreacted oxidizing agent is determined either by spectrometry or titration. The quantity of oxidizing agent used can be calculated and the oxygen equivalent then determined. Such methods include the following.

Measurement of the two-hour dichromate value. Here, the sample is refluxed with excess potassium dichromate in concentrated sulfuric acid for 2 h:

$$Cr_2O_7^{2-} + 14H^+ + 6e \longrightarrow 2Cr^{3+} + 7H_2O \tag{3.10}$$

Silver sulfate may be included to catalyse the oxidation processes of alcohols and low-molecular-weight acids.

Chloride ions give a positive interference by the reaction shown in the following equation. The interference is reduced by the addition of mercury (II) sulfate, with a chloro complex being formed:

$$Cr_2O_7^{2-} + 6Cl^- + 14H^+ \longrightarrow 2Cr^{3+} + 3Cl_2 + 7H_2O \tag{3.11}$$

If the excess dichromate is determined by titration, then iron (II) ammonium sulfate can be used:

$$6Fe^{2+} + Cr_2O_7^{2-} + 14H^+ \longrightarrow 6Fe^{3+} + 2Cr^{3+} + 7H_2O \tag{3.12}$$

DQ 3.9

Given that oxidation by oxygen can be represented by the following:

$$O_2 + 4H^+ + 4e \longrightarrow 2H_2O \qquad (3.13)$$

what is the equivalence between dichromate and oxygen?

Answer

One mole of $Cr_2O_7^{2-}$ consumes six moles of electrons to produce two moles of Cr^{3+}. Since each mole of O_2 can consume four moles of electrons to make H_2O, then one mole of $Cr_2O_7^{2-}$ is equivalent to 1.5 moles O_2.

In high-throughput laboratories, commercially available kits may be used which determine the unused dichromate by measuring the absorbance at 620 nm. Semi-micro systems are available which allow the test to be performed by using 2 ml aliquots of sample, heating in a sealed tube with premixed reagents. With such systems, 25 analyses can be performed simultaneously.

Permanganate tests. In these methods, excess potassium permanganate is added under specified conditions, which can range from three minutes on a steam bath to four hours at room temperature. The unreacted permanganate can be determined by any of a number of techniques, including the liberation of iodine, followed by titration of the latter with thiosulfate, as follows:

$$2MnO_4^- + 16H^+ + 10I^- \longrightarrow 2Mn^{2+} + 8H_2O + 5I_2 \qquad (3.14)$$

$$I_2 + 2S_2O_3^{2-} \longrightarrow S_4O_6^{2-} + 2I^- \qquad (3.15)$$

The confusing number of variations of this method leads to limitation in its use since it is very difficult to obtain comparative inter-laboratory data. The 'three-minute' variant of the test does, however, provide a rapid method of testing a specific water for its oxidizing ability.

A comparison of the BOD and COD tests is given in Table 3.2

DQ 3.10

On the basis of the comparisons given in Table 3.2, suggest appropriate applications of the two techniques.

Answer

BOD Long-term monitoring of natural water.
COD Rapid analysis of heavily polluted samples, e.g. industrial effluents.

Relationship of Oxygen Demand to Specific Concentrations. If a single organic compound was present in the water and the oxidation reactions proceeded

Table 3.2 Comparison of various BOD and COD tests

BOD	COD[a]
Five day analysis time	Rapid analysis
Closely related to natural processes	Less relationship to natural processes
Difficult to reproduce, both within laboratories and between laboratories	Good reproducibility
Care has to be taken with polluted water	Can analyse heavily polluted water

[a] All of these tests will be affected by the presence of inorganic reducing or oxidizing agents, with the former giving positive results, and the latter (possibly) leading to negative results.

to completion, the above methods would give an accurate measurement of its concentration. The determination of known amounts of a single compound can be used in the laboratory to test experimental procedures. Potassium hydrogenphthalate is often used, which is oxidized according to the following equation:

$$C_8H_5O_4K + 15/2O_2 \longrightarrow 8CO_2 + 2H_2O + K^+ + OH^- \qquad (3.16)$$

DQ 3.11

What is the COD of a solution containing 0.340 g l^{-1} potassium hydrogenphthalate?

Answer

The relative molecular mass of potassium hydrogenphthalate = 204.
In 1 l of solution, there are 0.340/204 mol.
1 mol potassium hydrogenphthalate = 7.5 mol oxygen.
Therefore, 0.340/204 mol potassium hydrogenphthalate

$$= 7.5 \times 0.340/204 \text{ mol oxygen}$$

$$= 7.5 \times 0.340/204 \times 32 \times 1000 \text{ mg oxygen}$$

$$= 400 \text{ mg}$$

Hence, the COD = 400 mg l^{-1}

3.3.3 Total Organic Carbon

None of the oxygen demand methods give a precise estimation of the total organic carbon (TOC) loading of the water. A number of techniques are available which can achieve this. All involve the oxidation of the organic matter to carbon dioxide, after prior acidification to remove interference from carbonates. The methods used include the following:

(i) Injection of a small quantity of water into a gas stream passing through a heated tube to carry out the oxidation. Measurement using this technique is possible to the mg l^{-1} level.
(ii) Wet oxidation by using potassium peroxydisulfate at room or elevated temperatures. This method is about 100 times more sensitive than the heated-tube oxidation approach.

The carbon dioxide can then be measured either by absorption in solution and measurement of its conductivity, by reduction to methane and analysis of this gas by flame ionization detection (see Section 4.2 below) or by direct measurement by infrared spectrometry (see Section 6.3 below).

Attempts are often made to replace BOD and other oxygen demand measurements with TOC. To understand this, you should note the following advantages:

(i) it is a rapid technique;
(ii) it would be expected to give highly reproducible results;
(iii) it can be easily automated, either for laboratory analysis or for on-line monitoring of effluents.

3.3.4 pH, Acidity and Alkalinity

The pH is related to the number of hydrogen ions in solution by the following relationship:

$$pH = -\log_{10} a(H^+) \tag{3.17}$$

where $a(H^+)$ is the hydrogen ion activity.

At the low concentrations of hydrogen ions and low ionic strengths which are typical of unpolluted environmental samples, the hydrogen ion activity is approximately equivalent to the hydrogen ion concentration. Some typical pH values found with environmental water samples are shown in Figure 3.8.

Did you realize that unpolluted rain water is slightly acidic? This is due to the presence of dissolved carbon dioxide, as follows:

$$\begin{aligned} H_2O + CO_2(\text{gas}) &\rightleftharpoons H_2O.CO_2(\text{solution}) \\ &\rightleftharpoons H^+ + HCO_3^- \rightleftharpoons 2H^+ + CO_3^{2-} \end{aligned} \tag{3.18}$$

Hardness in water is due to the presence of polyvalent metal ions, e.g. calcium and magnesium, arising from dissolution of minerals. For instance, the dissolution

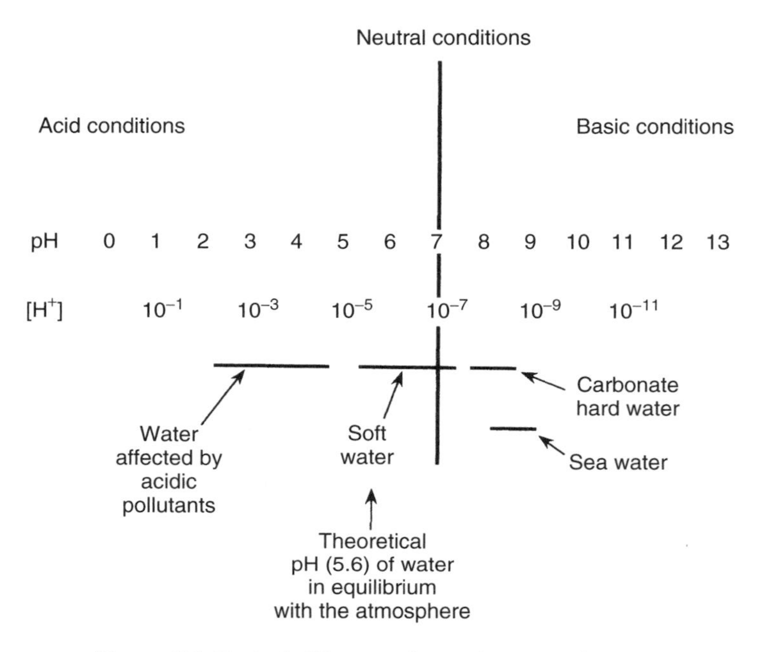

Figure 3.8 Typical pH ranges for environmental waters.

of limestone involves the equilibria shown in equations (3.19) and (3.20) below. From these equilibria, you should be able to see that the water will then be slightly alkaline:

$$CaCO_3 \rightleftharpoons Ca^{2+} + CO_3{}^{2-} \tag{3.19}$$

$$CO_3{}^{2-} + H_2O \rightleftharpoons HCO_3{}^{-} + OH^- \tag{3.20}$$

The biological effect of a change in pH can most easily be seen by the sensitivity of freshwater species to acid conditions. Populations of salmon start to decrease below pH 6.5, perch below pH 6.0, and eels below pH 5.5, with little life possible below pH 5.0. The eradication of life can result from a change of little more than 1 pH unit. Chemical effects are also observed. In Section 2.4 above, we discussed how a decrease in pH increases the solubility of metals. The use of lead piping for domestic water supplies becomes of greater concern as the water becomes more acidic. The weathering of minerals, such as limestone or dolomite, by water becomes more rapid with a decrease in pH.

A typical procedure for the measurement of pH involves calibration with two buffer solutions spanning the expected pH of the sample, followed by measurement of the sample.

The procedures for 'Alkalinity' and 'Acidity' measure, by titration, the quantity of acid or base needed, respectively, to change the pH of a sample to 4.5, corresponding to the methyl orange end-point. From a chemical point of view, this gives a measurement of the buffer capacity (resistance to change in pH) of the water. This resistance to change could be caused, for instance, by the presence of carbonate or hydrogencarbonate ions, as shown by equations (3.21) and (3.22) below. Indeed, the units of alkalinity and acidity are expressed as mg l^{-1} $CaCO_3$, regardless of the true species producing the effect:

$$CO_3{}^{2-} + H^+ \rightleftharpoons HCO_3{}^- \quad (3.21)$$

$$HCO_3{}^- + H^+ \rightleftharpoons H_2O.CO_2 \rightleftharpoons H_2O + CO_2 \quad (3.22)$$

A high buffer capacity is a useful feature if an acidic or basic pollutant is being added to water, as this will lessen the pH change of the receiving water.

3.3.5 Water Hardness

The term 'water hardness' will be very familiar to you if you live in an area where there are high concentrations of calcium and magnesium in your water supply. The effects you may have noticed include the following:

(i) Deposition of a white solid whenever the water is heated. This is commonly seen as the 'furring-up' of kettles. This may also lead to blockage of hot-water pipes and a decrease in the efficiency of industrial heat exchangers.

(ii) The formation of scum whenever soap or washing powder is added to water. Sometimes, coloured spots are produced on clothes. No detergent action can occur until all of the hardness has been removed from the water.

The effects are generally produced by the presence of polyvalent metal ions in the water from the weathering of minerals. Usually, this is almost entirely due to calcium and magnesium ions, although others such as aluminium, iron, manganese and zinc ions may make a small contribution. It is the transition metal ions which produce the staining often observed. The minerals producing the hardness are often based on carbonates (limestone ($CaCO_3$) and dolomite ($CaCO_3.MgCO_3$)) or sulfates (gypsum ($CaSO_4$)). It is only the hardness derived from carbonates which gives rise to solid deposition ('carbonate' hardness or 'temporary' hardness). Hardness which does not produce this effect is known as 'non-carbonate' or 'permanent' hardness.

However, a small degree of hardness does have some beneficial effects. For example, the alkalinity lowers the solubility of toxic metals, while the buffering action of the carbonate hardness lessens the effect of acidic pollutants. This buffer effect increases with the concentration of hardness in the water. In addition, there is evidence that hard water is beneficial to health, particularly in the reduction of heart disease, and it certainly is more pleasant to drink.

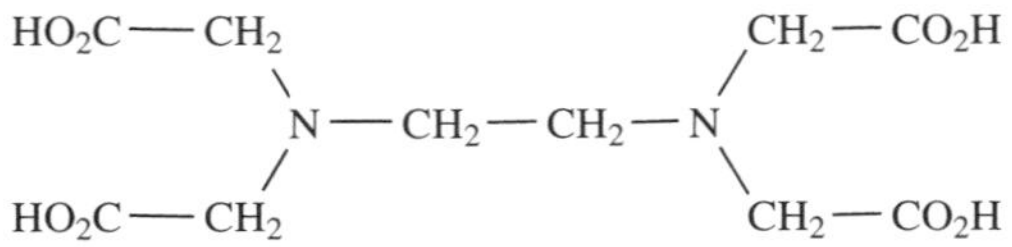

$$\begin{array}{ccc} HO_2C-CH_2 & & CH_2-CO_2H \\ \quad\diagdown & & \diagup \\ & N-CH_2-CH_2-N & \\ \quad\diagup & & \diagdown \\ HO_2C-CH_2 & & CH_2-CO_2H \end{array}$$

Figure 3.9 Structure of ethylenediaminetetraacetic acid.

Analysis is normally performed by complexometric titration using the disodium salt of ethylenediaminetetraacetic acid (Figure 3.9). This forms a 1:1 complex with divalent metal ions, according to the following:

$$M^{2+} + H_2EDTA^{2-} \rightleftharpoons M(EDTA)^{2-} + 2H^+ \qquad (3.23)$$

where H_2EDTA^{2-} is the di-anion derived from the acid

To determine both calcium and magnesium by titration, the pH has to be buffered at pH 10. The end-point is detected by using an indicator such as Erichrome Black T. A calcium-only value can be found by titrating at a higher pH. Under these conditions, the magnesium would precipitate as $Mg(OH)_2$.

The titration estimates the *total* divalent metal as a molar concentration. Many non-chemists are unfamiliar with molar concentrations and so the quantity is often re-expressed in more familiar terms. However, it would be impossible to convert the value into the more familiar weight concentrations (mg l^{-1}) without knowing the precise individual concentrations of calcium, magnesium and the other ions. Even then, you would not be able to quote a single figure for the total hardness – just a table of individual concentrations. In order to overcome this, the total hardness is expressed in mg l^{-1} units as if it were all calcium carbonate, even if it is due to calcium sulfate, magnesium carbonate or any other polyvalent metal salt.

DQ 3.12

Which of the following solutions give 50 mg l^{-1} total hardness?

(a) 50 mg l^{-1} $MgCO_3$
(b) 21.1 mg l^{-1} $MgCO_3$ + 25 mg l^{-1} $CaCO_3$
(c) 50 mg l^{-1} $CaSO_4$
(d) 55 mg l^{-1} $CaCl_2$

Answer

The concentrations of the above, expressed as molarities (M), are (a) 0.78, (b) 0.5 (0.25 + 0.25), (c) 0.37, and (d) 0.5 mM.

The hardness can be determined by multiplying by the relative molecular mass of calcium carbonate (= 100). This gives the total hardness of the solutions as (a) 78, (b) 50, (c) 37, and (d) 50 mg l^{-1}.

What values should we expect from environmental samples? Although the terms 'hard' and 'soft' sound very subjective, they have come to be defined within very specific concentration ranges. The definitions used within the United Kingdom are shown in Table 3.3.

DQ 3.13

Two unpolluted waters, which have the same pH value of 7.8, are contaminated by approximately the same quantity of acidic pollutant. The pH of one drops sharply, while there is only a small drop with the second. Suggest a reason for the difference and the analyses which could be used to confirm your suggestion.

Answer

The pH values would suggest hardness, perhaps from the presence of limestone (see Section 3.3.4). The hardness of the solutions could be determined by EDTA titration (see Section 3.3.5). The solution having the smallest pH change indicates a greater buffer capacity and so would be expected to have the greatest hardness.

3.3.6 Electrical Conductivity

You may sometimes wish to know the total inorganic salt content in a sample. A simple method would be to evaporate the sample to dryness and then weigh the resulting solid. Large volumes of sample would, however, need to be evaporated, thus making the technique less attractive than at first thought. It would be much more convenient if an electrode could simply be placed in the sample to

Table 3.3 Hardness descriptions used in the United Kingdom

Concentration (mg l^{-1} $CaCO_3$)	Description
0–50	Soft
50–100	Moderately soft
100–150	Slightly hard
150–200	Moderately hard
200–300	Hard
> 300	Very hard

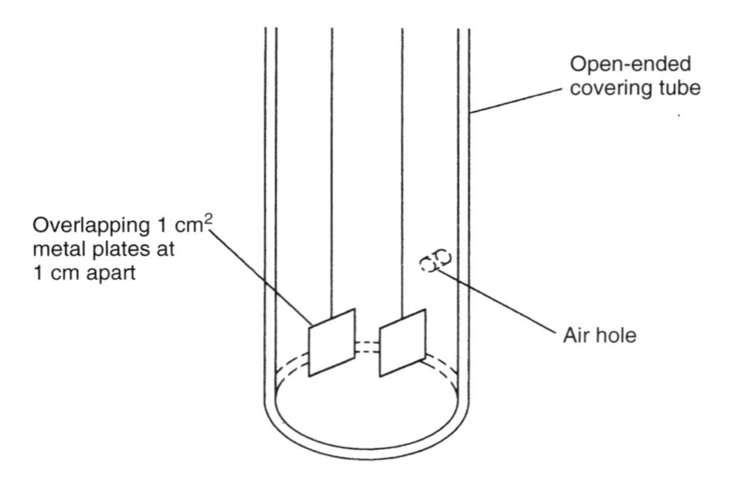

Figure 3.10 Schematic of a typical conductivity cell.

make the measurement. The closest method to this ideal situation is the use of a conductivity cell for dissolved ions, as illustrated in Figure 3.10.

Using this method, a low-voltage alternating current is applied across the electrodes. The resistance of the liquid between the electrodes is measured, and is converted to conductivity according to the following formula:

$$K = \frac{L}{AR} \tag{3.24}$$

where K is the conductivity, L the distance between the electrodes (cm), A the surface area of the electrodes (cm^2), and R the resistance (ohm = siemens $(S)^{-1}$) (note that the siemen is the SI unit of electric conductance).

The units of conductivity applicable to environmental samples are $\mu S\ cm^{-1}$, with a typical value of 200 $\mu S\ cm^{-1}$ being found for a soft water with a significant ionic salt content. The cell is calibrated by using solutions of known conductivity. Conductivity is highly temperature-dependent and so care has to be taken that calibration solutions and the unknown sample are at the same temperature. A standard temperature of 25°C is often used. The relationship between conductivity and total salt content is not simple. All ions having the same charge have approximately the same conductivity, but unfortunately most environmental waters contain ions with different charges in varying concentrations. If a series of waters of roughly similar composition is known, an approximate conversion can be made. For many waters in the United Kingdom, the following equation is valid:

$$\text{total salt concentration} = A \times \text{conductivity (mg l}^{-1}\text{)} \tag{3.25}$$

where A is a constant in the range 0.55 to 0.80.

SAQ 3.4

The UK General Quality Assessment (GQA) for rivers has the following limiting parameters:

GQA grade	Description	Dissolved oxygen (%)	BOD (mg l^{-1})	Ammonia (mg l^{-1}(N))
A	Very good	80	2.5	0.25
B	Good	70	4.0	0.6
C	Fairly good	60	6	1.3
D	Fair	50	8	2.5
E	Poor	20	15	9
F	Bad	< 20	> 15	< 9

Why do you consider that these parameters and limiting concentrations are used?

3.4 Techniques for the Analysis of Common Ions

This section discusses the application of techniques to quantify ions present in the mg l^{-1} range. Revise Section 3.1 if you don't remember which these are. You will notice that from this section onwards the techniques are almost entirely instrumental, thus confirming that one of the major advantages of such methods is to be able to analyse low concentrations of multi-analyte samples with ease. Many of the instrumental methods for ions within the mg l^{-1} concentration range need little sample preparation. Later, when we look at the analysis of ions at μg l^{-1} levels, much of the discussion will be concerning preconcentration to bring the samples to within the working range of the instruments. The instrumental method then becomes just one part of a more complex analytical procedure.

3.4.1 Ultraviolet and Visible Spectrometry

DQ 3.14

From your knowledge of this spectroscopic technique, describe the law on which the analytical method is based.

Answer

At sufficiently low concentrations, the Beer–Lambert law is followed:

$$A = \varepsilon cl \qquad (3.26)$$

where A *is the absorbance of radiation at a particular wavelength* (= *log*(I_0/I)), I_0 *the intensity of the incident radiation,* I *the intensity of the* *transmitted radiation, ε the proportionality constant (molar absorptivity ($l\ mol^{-1}\ cm^{-1}$)), c the concentration of the absorbing species ($mol\ l^{-1}$), and l the pathlength of the light-beam (cm).*

If you had difficulty in remembering this law, then it would be useful to carry out some revision before proceeding any further. The Beer–Lambert law is fundamental to many of the techniques that we will be discussing in the following sections. The instruments used to measure the absorption of light can range from sophisticated laboratory instruments which can operate over the whole ultraviolet/visible range to portable colorimeters employing natural visible light, which are used as field instruments. This makes absorption spectrometry one of the most useful and versatile techniques for an environmental analyst.

You might at first hesitate to believe this last statement. After all, none of the common ions in water absorb light in the visible region of the spectrum. You know this because natural water is usually almost colourless. In addition, the only ions commonly found in water which absorb in the ultraviolet range above 200 nm are nitrate and nitrite. The main use of the technique involves the analysis of light-absorbing derivatives of these ions. This can be carried out for almost all of the common anions (except sulfate), as well as ammonia. We can summarize as follows:

Analysis by direct absorption
nitrate

Analysis after formation of derivative

chloride	fluoride
nitrate	nitrite
phosphate	

As an example of such an approach, the procedure for phosphate involves the addition of a mixed reagent (sulfuric acid and ammonium molybdate, ascorbic acid and antimony potassium tartrate) to a known volume of sample, making up to the working volume, shaking and leaving for 10 min. A blue-coloured phospho-molybdenum complex is produced, and its absorbance is measured at 725 nm.

The chemistry behind the colour-forming reactions has been long established and is well understood.

3.4.1.1 Quantification

Ultraviolet/visible spectrometry is the first technique we have discussed where, at low concentrations, there is a simple linear correlation between the instrument response and the concentration of the unknown.

DQ 3.15

How would you set about using this technique for a quantitative analysis?

Answer

You would make up a series of standard solutions of known concentration of the unknown and from this construct a calibration curve. The concentration should be within the range over which the Beer–Lambert law applies and thus a straight-line graph will be produced. Above this range, the calibration will no longer be linear and the solutions should be diluted. The 'best-fit' calibration line can readily be calculated by using the method of least-squares found on standard PC spreadsheets or even the most basic scientific calculators. The absorbance of the unknown can then be measured and from this the concentration calculated.

This procedure is known as *calibration by external standards*. We will find instances in the following sections where the sample matrix can affect the response of the instrument and so 'external standards' may not be the best method to employ.

DQ 3.16

Plot a calibration graph from the following data and determine the concentration of phosphorus in the sample:

Concentration ($\mu g\ l^{-1}$(P))	25	50	125	250	375	Unknown
Absorbance	0.058	0.149	0.370	0.683	1.060	0.426

Answer

The least squares line is:

$$Absorbance = 0.002\,81\,(concentration) + 0.001\,04$$

which gives the unknown concentration as 151 $\mu g\ l^{-1}$ *(P).*

3.4.1.2 High-throughput Laboratories

You will find a great diversity of instrumentation based on these chemistries for high-throughput laboratory analysis. A number of instruments are based on continuous flow, with a schematic of a typical system being shown in Figure 3.11. Instead of prior mixing of the reagents for each analysis, streams of each reagent (segmented by air bubbles to diminish premature mixing effects) in narrow-bore tubes are mixed by combining the flows at a T-junction or within a (mixing) cell. A sample is introduced from an automatic sampler as a continuous flow into the reaction stream. The combined flow is then led into a spectrophotometer and the absorption measured. The flows of all of the reagents and samples are controlled from a multi-channel peristaltic pump (Figure 3.12.).

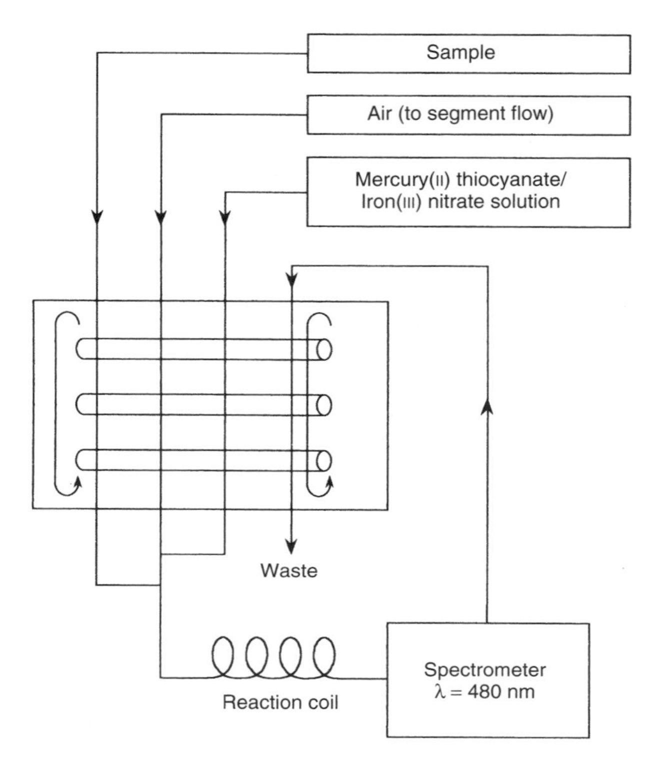

Figure 3.11 Schematic of a typical continuous-flow system used for the analysis of chloride ions.

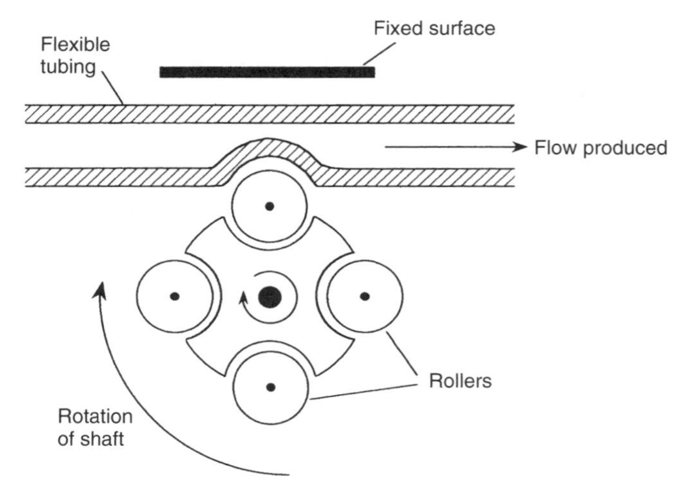

Figure 3.12 Schematic of the operation of a peristaltic pump.

Other instruments are based on flow-injection techniques where individual aliquots of sample are injected into a continuous flow of water. Colour-forming reagents are then added, also via a continuous-flow system. The mixing of the reagents and samples is dependent on the length and diameter of the tubing. After time being allowed for the formation of the colour, the absorbance of the solution at a specific wavelength is then measured. The response of the instrument is in the form of a peak, with the peak height being proportional to the sample concentration (Figure 3.13).

If there is a requirement to analyse more than one of the ions, then discrete analysers may be used. In such systems, the samples are introduced into vials on a rotating carousel. As the carousel rotates, reagents are added and mixed, time is allowed for colour development, and the light absorbance at a specific wavelength is then determined. If the instrument is suitably configured, several different analyses can be performed simultaneously on samples by one instrument.

3.4.1.3 Field Techniques

Field techniques are becoming increasingly important for giving immediate measurement of ion concentrations. Un-manned field stations can be set up by using the automatic procedures described above. Alternatively, portable (often hand-held) instruments may be used.

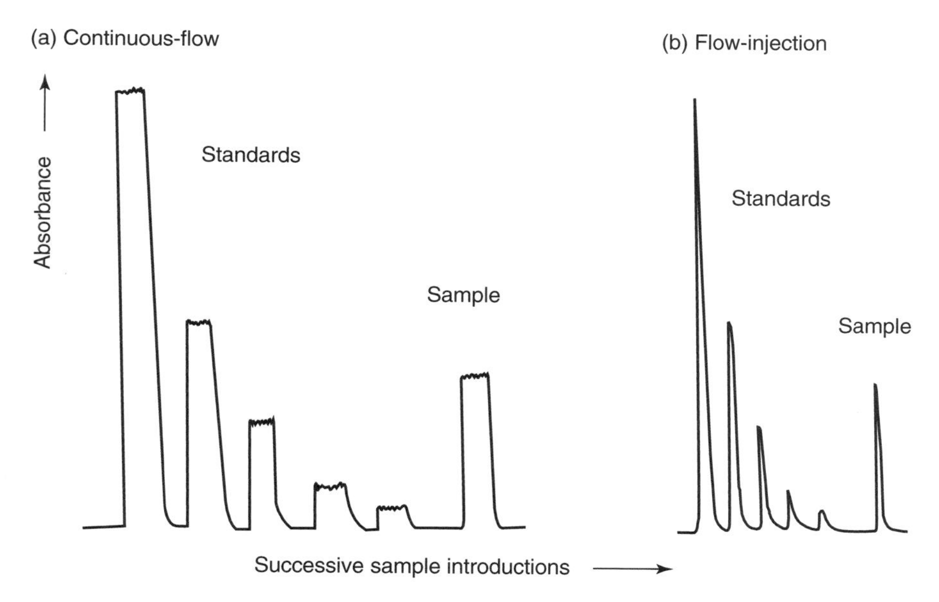

Figure 3.13 Typical outputs obtained from (a) continuous-flow and (b) flow-injection analysers.

DQ 3.17

What modifications need to be made to the standard apparatus and methods already described for use in portable field instruments?

Answer

The procedure for the colour-forming reaction has to be made simple. No one wishes to perform complicated analytical routines on a muddy riverbank! Calibration of the instrument should avoid the use of standard solutions, which, once again, are inconvenient in the field. The optical components of the instrument should be minimized or, at the very least, be made robust.

Each manufacturer has a different approach to such modifications. Colour-forming reagents may be pre-measured in the form of tablets, or in solution. As a further simplification, one manufacturer seals the reagents under vacuum in an ampoule. Breakage of the top under water automatically draws the correct sample volume into the ampoule. Coloured glass or moulded plastic standards are often used rather than solutions. These can be in the form of a disc. One manufacturer's design contains glasses of different optical density (Figure 3.14). The disc is rotated through the light beam until the colour of the standard glass matches that of the unknown. Alternatively, a moulded plastic cube may be used which has a stepped side to provide a number of possible pathlengths (and hence

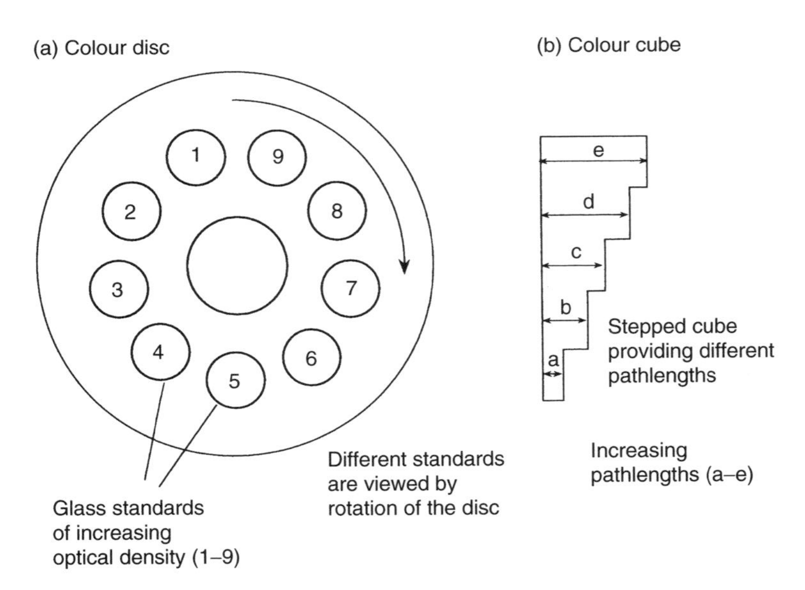

Figure 3.14 Typical designs of colour standards: (a) colour disc; (b) colour cube.

absorbances). The most simple procedure for quantification is by visual comparison of the colour of the standards and the unknown using available sunlight. Alternatively, portable spectrometers are available, often housed in briefcases, along with titration equipment and pH and conductivity electrodes – these are known as 'water quality' test kits.

3.4.1.4 A Note on Units

The most obvious way of expressing the concentration of ions is as the mass of the ion per unit volume (mg l^{-1}), but sometimes you will find other units, most notably as the concentration of the major element within the ion. This alternative method is most common for the nitrogen-containing ions. Nitrate, nitrite, and ammonium are often expressed as mg l^{-1} of NO_3^-, NO_2^- and NH_4^+, respectively, but can all be expressed as mg l^{-1} of nitrogen (mg l^{-1}N). It then becomes easy to compare the relative concentrations of species without having to use molarities. If all of the ammonia in a water sample which contains a concentration of 2 mg l^{-1} (expressed as nitrogen) is totally converted into nitrate, then the water will contain a nitrate concentration of 2 mg l^{-1} (also expressed as nitrogen). This is easier for a non-specialist to understand than by saying that 3.09 mg l^{-1} NH_4^+ will produce 8.86 mg l^{-1} NO_3^-. Difficulties can arise because the two systems are sometimes used in parallel. For instance, the UK uses nitrogen (N) concentrations (look back at SAQ 3.4), whereas the European Union (EU) legislation is based on expressing the concentrations as the ions (look back at Table 1.2).

A similar difficulty arises for phosphorus ions, where in the UK they are expressed as mg l^{-1} P rather than as individual ion concentrations. In this case, the EU system expresses the concentrations as P_2O_5, with these somewhat unexpected units having been long used in agriculture!

DQ 3.18

What would be the concentration of the following?

(i) 50 mg l^{-1} (NO_3^-) expressed as mg l^{-1} (N).
(ii) 100 mg l^{-1} (P) expressed as mg l^{-1}(P_2O_5) and as mg l^{-1} (PO_4^{3-}).

Answer

(i) 50 mg l^{-1} NO_3^- is equivalent to 50/62 mol l^{-1} nitrate
= (50 × 14)/62 mg l^{-1} nitrogen
= 11.3 mg l^{-1} N

(ii) 100 mg l^{-1} P is equivalent to 100/31 mol l^{-1} phosphorus
= 100/(31 × 2) mol l^{-1} P_2
The relative molecular mass of $P_2O_5 = (2 \times 31) + (5 \times 16) = 142$
and the relative molecular mass of $PO_4^{3-} = 31 + (4 \times 16) = 95$

Hence, $100\ mg\ l^{-1}\ P = (100 \times 142)/(31 \times 2)\ mg\ l^{-1}\ P_2O_5$
$= 229\ mg\ l^{-1}\ P_2O_5$
Similarly $100\ mg\ l^{-1}\ P = (100 \times 95)/31\ mg\ l^{-1}\ PO_4{}^{3-}$
$= 306\ mg\ l^{-1}\ PO_4{}^{3-}$

3.4.2 Emission Spectrometry (Flame Photometry)

Emission spectrometry relies on the principle that, for some metals at low concentrations, the intensity of light emitted from an electronically excited atom (usually produced by introduction of the sample into a flame) is proportional to the concentration of the excited species. Simple and inexpensive instrumentation is available, often known as 'Flame Photometers'.

DQ 3.19

Flame photometry seems almost ideally suited to the analysis of environmental water samples. List some of the reasons for this from your prior knowledge of the technique.

Answer

1. *Although the use of flame photometry is limited to a few alkali metal and alkaline-earth ions, this includes sodium, potassium and calcium, three of the four major cations present in water. Check in Section 3.1 earlier if you are unsure of the fourth.*
2. *The linear concentration ranges (0–10 mg* l^{-1} *(for sodium and potassium), and 0–50 mg* l^{-1} *(for calcium) are within that expected for environmental water samples (see Section 3.1 above). Little sample preparation is needed.*
3. *The instrument is simple to use and the only laboratory requirements are a gas supply (natural gas is adequate) and a source of vacuum. This can be easily installed in temporary laboratories for analysis close to the sampling site.*

It is a pity that flame photometers cannot be used to analyse the fourth common ion, i.e. magnesium, as all of the routine analytical requirements for metal ions could then be satisfied by this simple method. Analysis for magnesium is usually carried out by using atomic spectrometry (see Section 4.3 below)

The major disadvantage of flame photometry is the variation of the response of the instrument with time (i.e. *drift*). Great care has to be taken to ensure that calibration of the instrument and the analytical measurements are performed quickly after each other. It is also good practice to repeat the calibration after the analysis to check that no variation has occurred.

3.4.3 Ion Chromatography

The methods we have looked at so far have been for the analysis of individual ions, but sometimes a complete analysis of all of the ions in the sample is needed. Chromatographic separation of the ions is an obvious approach. Liquid chromatography would seem particularly useful since the species to be analysed are already in solution. From your reading elsewhere, you will be familiar with the principles of high performance liquid chromatography (HPLC) and how its application over the last three decades has expanded to include virtually all soluble ions and compounds. The major application in environmental analysis has been for inorganic anions. Several variations of the liquid chromatographic technique have been developed which normally use specialized 'ion chromatographs'. The most sensitive systems are often those which use a technique known as ion suppression (Figure 3.15).

The separation of the anions is achieved by using an ion-exchange column (length 10–25 cm, 3–4.6 mm i.d.), usually based on poly(styrene–divinylbenzene) or another organic polymer, with an eluent typically containing sodium hydroxide or a sodium carbonate/hydrogencarbonate buffer. Detection of the analyte ions is achieved by monitoring the increase in conductivity of the eluent produced by the ions as they pass through the detector. In order to maximize detection sensitivity, prior to passing to the detector, all buffer ions have to be removed from the eluent as these would contribute to the background conductivity. The sodium ions in solution are replaced by hydrogen ions. The hydroxide ions react to form water (see equation (3.27)). Carbonate and hydrogencarbonate react to form carbon dioxide (see equations (3.28) and (3.29)), which has little conductivity in solution.

Hydroxide eluents:
$$OH^- + H^+ \rightleftharpoons H_2O \tag{3.27}$$

Carbonate/hydrogencarbonate eluents:
$$HCO_3^- + H^+ \rightleftharpoons H_2O + CO_2 \tag{3.28}$$
$$CO_3^{2-} + 2H^+ \rightleftharpoons H_2O + CO_2 \tag{3.29}$$

The suppressor has to provide, uninterruptedly, precisely the correct number of protons for the neutralization. There are a number of methods used to achieve this. All of these are based on the ion-exchange process.

One manufacturer uses a continuous suppression system, as shown in Figure 3.16. The eluent passes between cation-exchange membranes, through the

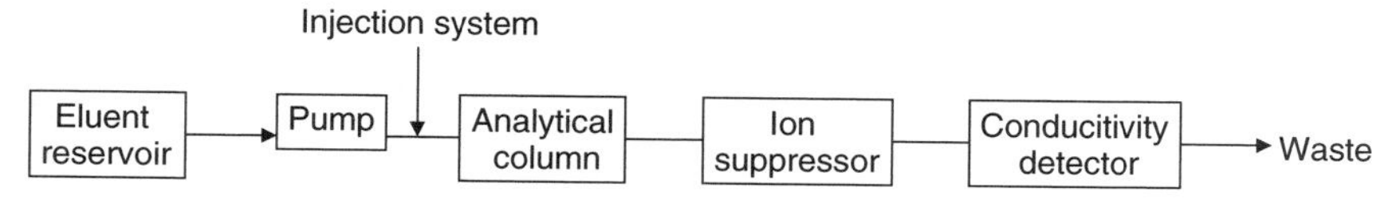

Figure 3.15 Major components of a suppressed ion chromatographic system.

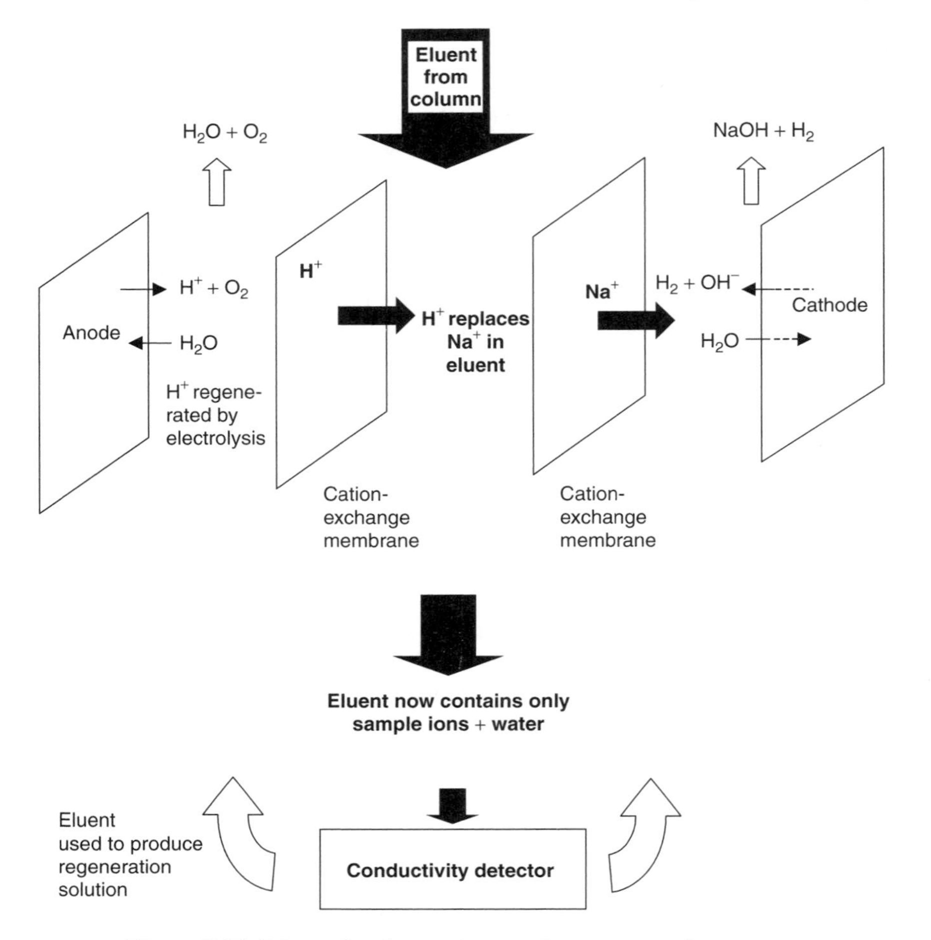

Figure 3.16 Schematic of a continuous eluent suppression system.

detector cell, and is finally recycled on the outside of the membranes. The H^+ ions necessary to replace the Na^+ ions in the fresh eluent are generated by electrolysis of the recycled eluent, with the H^+ ions being generated at the cathode.

Other manufacturers use cation-exchange columns which need periodic regeneration. This can be achieved without interruption of the analytical operation of the chromatograph. In one system, there is a carousel of regeneration columns with one column being regenerated while another one is in use. Another system regenerates the column while the following sample is being loaded. Disposable regeneration columns may also be used.

The response plotted in the chromatogram is *conductivity*. As the latter is directly proportional to the concentration of the ion, quantification can be simply carried out by comparison of the peak area of the unknown with that of a standard of similar concentration, i.e. by external standards.

For ions of interest in environmental water found at mg l^{-1} concentrations and with a suppressed system the sample would need to be diluted before injection. This, along with filtration, is often the only sample preparation necessary and common ions in water can be determined within the space of a few minutes (Figure 3.17).

Non-suppressed ion chromatographs monitor the conductivity of the eluent directly, i.e. without the suppressor. Although the sensitivity is lower than that of suppressed systems, it is still sufficient to determine ions at mg l^{-1} concentrations. Non-suppressed systems have the advantage of being less complex instruments than suppressed chromatographs. Separation columns can be used with a wider range of eluents. The instruments resemble conventional liquid chromatographs (the components being simply pump, analytical column and detector) but often they contain no metal components in contact with the eluent, and with the pumps

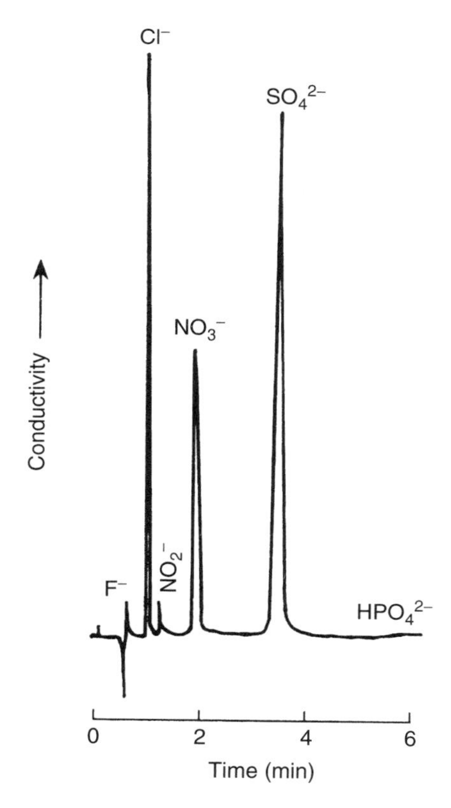

Figure 3.17 Typical chromatogram of a natural water sample.

designed to operate at lower pressures than is necessary for conventional HPLC. This reflects the lower back pressures found with ion chromatography when compared with reversed-phase liquid chromatography.

DQ 3.20

What disadvantage can you see in using ion chromatography for a single ion such as chloride or nitrate?

Answer

The time taken for the analysis will be determined by the elution time of the slowest component (often sulfate or phosphate), rather than the component of interest. A second analysis cannot proceed until all of the ions have eluted. This makes the technique slow in comparison to the methods dedicated to single ions, e.g. continuous-flow or flow-injection analysis.

Although most analysis nowadays would use specialized 'ion chromatographs', conventional high performance liquid chromatography may still find some application. Methods developed for conventional HPLC can use either a reversed-phase column and ion-pair techniques, or an ion-exchange column. Ultraviolet absorbance and conductivity detectors are used.

When a conductivity detector is employed, the system becomes similar to the specialized chromatographic set-up without ion suppression which has already been described. The sensitivity is lower in comparison with the specialized system, although it is still sufficiently high to analyse common anions at mg l^{-1} concentrations.

DQ 3.21

What common anions can be detected by using UV absorbance?

Answer

Nitrate and nitrite ions absorb in the UV range of the spectrum (see Section 3.4.1 above).

HPLC with UV detection is a useful method for these two ions when analyses of other ions are not required.

Although the most common use of chromatography is for anions, similar methods have been developed for specialized 'ion chromatographs' for the separation of the common cations (Na^+, K^+, $NH_4{}^+$, Ca^{2+}, Mg^{2+}, etc.) in a single isocratic run. A typical eluent would be methanesulfonic acid. This would allow ion suppression similar to that used for anions, although this may not be necessary at typical natural water concentrations.

3.4.4 Examples of the Use of Other Techniques

We have now discussed the most widely used methods for analysis of the common ions. There are, however, a few frequently used techniques which have not yet been covered. We will look at the analysis of three species – ammonia, fluoride, and sulfate – to exemplify these techniques.

3.4.4.1 Ammonia

Ammonia is the only alkaline gas commonly found in environmental water. If extracted from the sample, the ammonia can be determined by a simple acid–base titration. Magnesium oxide is added to the sample to make it slightly alkaline. The ammonia is then present predominantly in the form of NH_3, rather than the less volatile NH_4^+. Ammonia is then distilled off (Figure 3.18) and absorbed into boric acid solution. The boric acid sharpens the end-point of the subsequent titration with standard acid.

For rapid screening of samples, it is possible to use an ion-selective electrode (i.e. an electrode whose potential, measured with respect to a reference, is

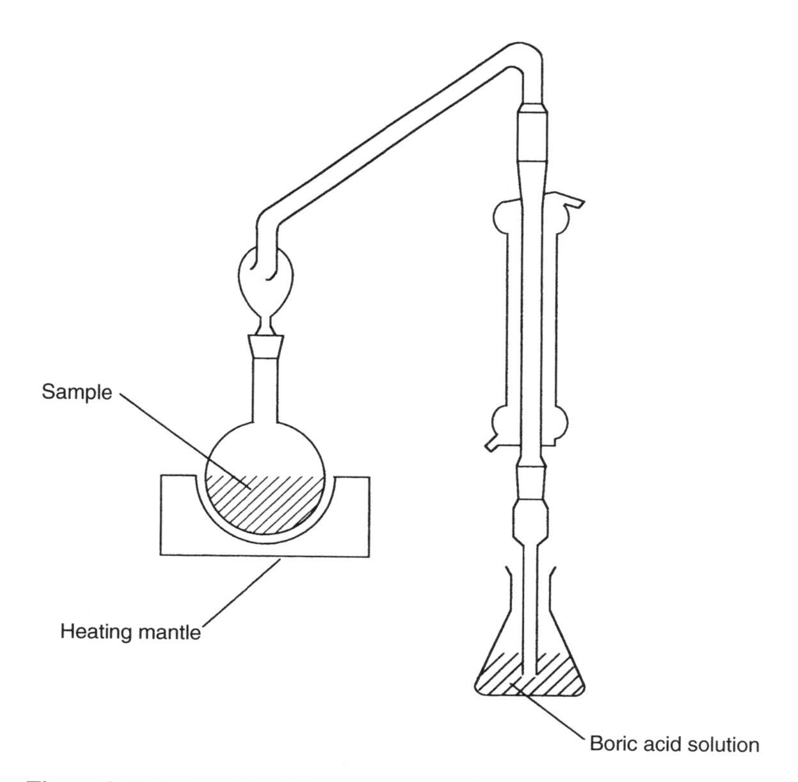

Figure 3.18 Schematic of the apparatus used for ammonia determination.

proportional to the log of the activity of one particular ion). Although this may appear to be a new technique for you to learn, you are already familiar with one particular ion-selective electrode. A combination pH electrode is simply an ion-selective electrode responsive to hydrogen ions and a reference electrode housed in a single body. Ion-selective electrodes are available for most common ions and gases which dissolve as ionic species, although they do have some limitations. Many are prone to interference from other species and thus have poor precision. Even the pH electrode has taken many years of development to produce reliable responses. All of them respond to ionic activity rather than concentration, and so it is essential to add a large excess of an ionic salt to both the standard solutions and the unknown in order that the ionic strength of each solution is identical.

Ammonia electrodes are of the gas-sensing type. The ammonia diffuses through a permeable membrane and causes a pH change in a small volume of internal solution, which is sensed by a 'glass' electrode. Prior to measurement, concentrated sodium hydroxide solution is added to the samples and standards. This serves to increase the pH to above 11 to ensure that the ammonia is in the unprotonated form, and also to provide a constant ionic strength. The ammonia electrodes respond only to gaseous alkaline gases. For most environmental applications (except in the analysis of heavily polluted water), there will then be little possibility of interference. Calibration is by external standards.

You may wish to compare and contrast these methods with the spectrometric method (see Section 3.4.1 above) which can also be used for ammonia.

3.4.4.2 Fluoride

A second electrode which has found widespread use for water analysis is that which detects fluoride ions. This is a solid-state electrode where the electrical potential is generated by migration of the ion through a doped lanthanum fluoride crystal. This once again gives extremely high specificity to the analyte ion, with the only pretreatment necessary being the addition of buffer solution to maintain constant pH and ionic strength. Alternative techniques for fluoride determination are spectrometry (see Section 3.4.1.) and ion chromatography (see Section 3.4.3.).

3.4.4.3 Sulfate

There is no direct colorimetric method available for sulfate and ion-selective electrodes for the ion are not very reliable; in fact, the only direct instrumental method is by using ion chromatography. Virtually every other method is based on precipitation of an insoluble sulfate. Barium or 2-aminoperimidinium (Figure 3.19) salts are used for the precipitation. The precipitate formed may then be weighed for a direct determination of the sulfate. This represents one of the few remaining important applications of gravimetric analysis.

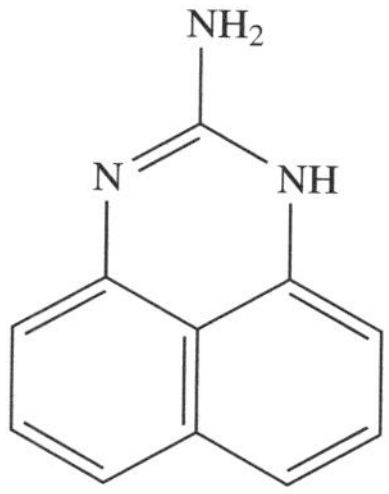

Figure 3.19 Structure of 2-aminoperimidine.

Other methods using insoluble salt precipitation are indirect, estimating the excess cation after precipitation of the sulfate. Excess barium may be determined by titration (which titration have you already come across which will analyse a divalent metal ion?) or by atomic absorption spectrometry (see Section 4.3.3 below) Excess 2-aminoperimidinium ions may be estimated by visible spectrometry.

None of these methods would appear ideal for a high-throughput laboratory. For most samples, sulfate would be the only major sulfur-containing species. Total sulfur in solution, as determined by an elemental sulfur analyser, will then give a good estimate of the sulfate concentration.

SAQ 3.5

Many of the species we have discussed can be analysed by more than one method. Tabulate the common procedures available for each of the major ions found in water.

SAQ 3.6

Consider the techniques you have listed in SAQ 3.5. What criteria would influence your choice of method?

Summary

The composition of water continuously changes as it travels in the environment. Sampling at a large number of locations is therefore necessary to monitor these changes. Careful choices of locations, sampling time, and sample storage procedures are necessary for reliable monitoring. The quality of water can be assessed by using measurements relating to the overall effects of groups of compounds or ions (water quality parameters), as well as by analysis of the major individual components. Methods for both types of determination have been described. These include both volumetric and instrumental methods.

Chapter 4

Water Analysis – Trace Pollutants

Learning Objectives

- To understand the need for extraction and pretreatment in the analysis of trace water components.
- To chose and apply suitable analytical methods for organic trace pollutants in water.
- To chose and apply appropriate methods for trace metal analysis in water samples.
- To understand what is meant by the term'speciation' and describe how it may be investigated for metals in water.

4.1 Introduction

Before you started this book, you may have thought that compounds with a concentration in water in the $\mu g\ l^{-1}$ range would have been of little environmental consequence. The introductory chapters showed how some ions and compounds could have effects significantly greater than what may have initially been expected from their low environmental concentrations. The two major groups are neutral organic compounds and some metal ions. These materials readily bioaccumulate and thus are found in organisms at concentrations exceeding the background levels by many factors of ten.

Another major cause of concern is the presence in water of a number of non-bioaccumulative organic compounds with adverse toxicological properties. For many years, there has been much concern over compounds suspected of being carcinogens. A typical example would be chloroform which can be produced in trace quantities during the disinfection of water by chlorination and which is thought to be harmful at $\mu g\ l^{-1}$ concentrations. Of more recent concern is the large number of compound types considered to be endocrine disruptors (see Sections 1.4 and 2.3 earlier). These compounds can range from pesticides, through components of common plastics, to active ingredients in the contraceptive pill.

In the early days of instrumental analysis the concentrations would have been beyond the capabilities of the available instrumentation and techniques but developments since then have made such analyses routine. This is partly due to the development of more sensitive instrumentation, but also through the development of suitable pretreatment processes. This is required to remove potential interferences and, for many techniques, to increase the analyte concentration to within the instrument sensitivity.

4.2 Organic Trace Pollutants

The range of organic compounds which may be found in environmental waters includes the following:

- Naturally occurring compounds from decaying organic material
- Pollutants discharged or escaping into the environment
- Degradation and inter-reaction products of the pollutants
- Substances introduced during sewage treatment

Typical analyses could include:

(i) Analysis of individual compounds or groups of compounds of environmental concern.

(ii) Total analysis of all organic components above the limit of detection. This is an enormous task and at the lower end of the concentration range there will almost invariably be unidentified components.

(iii) Field screening for specific pollutants prior to laboratory analysis.

(iv) Qualitative identification of trade products in spillages or discharges.

DQ 4.1

We found in earlier chapters that the properties of compounds causing widespread environmental problems include toxicity, slow biodegradation and the ability to bioaccumulate within organisms. List the types of organic compound which may be included in the classification.

Answer

You should have included the following in your list:

- *Pesticides, particularly those containing chlorine*
- *Chlorinated solvents*

- *Polychlorinated biphenyls*
- *Dioxins*
- *Endocrine disruptors*

For more localized pollution problems, we could extend our list of concern to include virtually every organic compound currently in use or production, together with their reaction and degradation products.

For the purpose of grouping into suitable analytical techniques, organic pollutants are often classified as being either 'volatile' (e.g. chloroform) and semi-volatile (e.g. most pesticides). The two groups may have different extraction and clean-up methods.

DQ 4.2

Analysis of complex mixtures of organics would normally involve the chromatographic separation of the components. Which form of chromatography would you consider most appropriate?

Answer

As most organic compounds have significant volatilities even at room temperature, gas chromatography would be expected to be a useful technique. The alternative of high performance liquid chromatography is used only where there are advantages over established gas chromatographic methods, although the number of applications of this technique is increasing.

A major area where non-chromatographic methods are used is in the determination of groups of compounds such as phenols, and also of classes of detergents, where the total concentration of the group of substances is required rather than the concentration of individual compounds.

DQ 4.3

What technique have you met which could analyse groups of organic compounds?

Answer

Ultraviolet/visible absorption spectrometry appears ideal. Absorptions are broad and the molar absorptivities often vary little between compounds within groups. A single absorption measurement could be used to determine the total concentration of the group. Although there may be suitable volumetric techniques for individual groups of compounds, they would not be sufficiently sensitive for concentrations in the $\mu g\ l^{-1}$ *range.*

The current desire for field screening has lead to novel approaches which may not have widespread use in other areas of chemical analysis. These include the use of immunoassays. After dealing with sample storage and extraction, this section will then look at gas chromatographic methods and later discuss the other techniques.

4.2.1 Guidelines for Storage of Samples and their Subsequent Analysis

DQ 4.4

In the last chapter, we covered general principles for sample storage. List the considerations necessary for organic trace pollutants.

Answer

The following list should not contain too many surprises:

(a) The volatility of organic compounds.
Even high-relative-molecular-mass compounds (e.g. pesticides) have a significant vapour pressure at room temperature. Storage containers should be completely filled and kept at sub-ambient temperatures; 4°C is often specified in analytical procedures. The latter is the temperature of a normal domestic refrigerator.

(b) Microbial degradation.
Storage at 4°C will lower microbial activity; storage below 0°C (i.e. deep freeze) will lower this still further.

(c) Photolytic decomposition.
Many potential analytes (e.g. organochlorine pesticides) are photosensitive in dilute aqueous solution. Therefore, the samples should be stored in the dark.

(d) Contamination from the container.
Glass bottles should be used, as bottles made of organic polymers will leach potentially interfering monomers and additives into the sample.

(e) Loss of Analyte on to container walls.
Low-solubility organic compounds can be adsorbed on to the container walls. This problem cannot be fully overcome. The best method of minimizing the effect is to proceed with the analysis as quickly as possible. Many procedures specify a maximum storage time.

The sample volumes which are required depend on the concentration of the analyte. Although the chromatographic techniques used involve the injection of just a few microlitres of solution, and spectrophotometric analysis a few millilitres, the solutions may first have been extracted from several litres of sample.

The precautions necessary to avoid either contamination or loss of material at these low concentrations, during the subsequent analysis, are often not appreciated. A few typical precautions should indicate the caution that is necessary:

(i) The analysis should be performed in a laboratory as free as possible from the analyte. Remember that many of these trace contaminants are solvents frequently found in analytical laboratories.
(ii) Any stock solvents should be safeguarded, minimizing exposure to the atmosphere and avoiding sample withdrawal with potentially contaminated pipettes or syringes.
(iii) Samples and working standards should be placed well away from more concentrated solutions or stock solvents.
(iv) As traces of pesticides are commonly found in laboratory solvents, pesticide-free grade solvents should be used for these analyses.
(v) Glassware should be scrupulously cleaned or new, if at all possible.

Such is the problem of contamination that the practical lower limits of detection can often be limited by the background concentrations of the analyte (or of interfering components) in the reagents or laboratory atmosphere.

4.2.2 Extraction Techniques for Chromatographic Analysis

Extraction of the compound of interest from the aqueous sample into an organic solvent is commonplace before any chromatographic analysis. The major reasons for this are as follows, while further advantages will be discussed below in Section 4.2.3:

- to separate unwanted components present in large excess
- to separate minor components which have overlapping peaks with the components of interest
- to concentrate the components of interest

For some samples, the extraction may be the only pretreatment necessary before injection into the chromatograph while for more complex samples it may be just one stage of a multi-stage process. Most of the techniques described may be integrated with the chromatographic stage and subsequent data handling. There is no one method of choice. The best method will be dependent on the following:

(i) The chemical and physical properties of the compounds being determined and potential interferences.

(ii) The choice of gas or liquid chromatography as the separative technique.

(iii) Whether solvent-free methods are preferred. Such methods remove the concern of possible contamination of the laboratory and its atmosphere (health effects and cross-contamination of other samples), contamination of aqueous waste and the cost of disposal of the waste solvent.

(iv) The number of samples to be analysed. If you have a large number of samples and are working in a well-equipped laboratory, the techniques which are fully integrated with the chromatograph may be preferable. In a smaller laboratory, dedicated instruments may not be justifiable and simpler methods may be preferred.

(v) Whether you would wish to perform the field extractions.

The extraction methods are common in many areas of chemical analysis. Try thinking of methods you have already come across before studying the following sections. These methods are summarized in Figure 4.1.

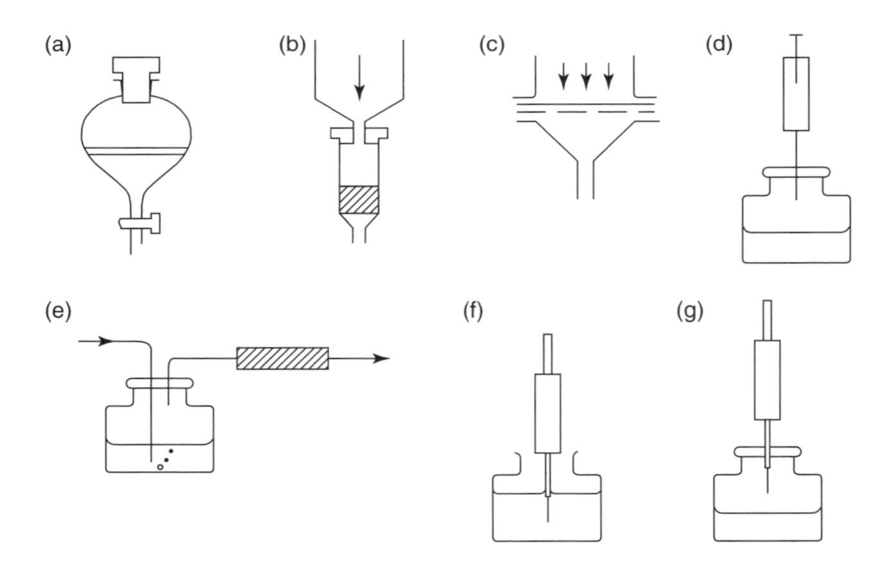

Figure 4.1 Summary of extraction methods: (a) solvent extraction; (b) solid-phase extraction – cartridge; (c) solid-phase extraction – disc; (d) head-space analysis; (e) purge and trap; (f) solid-phase microextraction – direct; (g) solid-phase microextraction – head-space.

4.2.2.1 Solvent Extraction

In this method, the water sample is shaken with an immiscible organic solvent in which the components are soluble. Hexane and petroleum ether are the most common extraction solvents, although oxygenated and chlorinated solvents are sometimes used. The organic layer is separated and, after drying, is injected into the chromatograph. The extractions can be made selective towards acidic and basic components by altering the pH of the aqueous layer. If the sample is acidified, the basic components are less likely to be extracted, for example:

$$\underset{\text{amine, soluble in non-polar solvents}}{RNH_2 + HCl} \longrightarrow \underset{\text{amine hydrochloride, less soluble in non-polar solvents}}{RNH_3^{+}\ Cl^{-}} \quad (4.1)$$

Similarly, if the sample is made basic, acidic components are less likely to be extracted, for example:

$$\underset{\text{carboxylic acid, soluble in non-polar solvents}}{RCO_2H + NaOH} \longrightarrow \underset{\text{carboxylate salt, less soluble in non-polar solvents}}{RCO_2^{-}\ Na^{+}} \quad (4.2)$$

When making the choice of extraction solvent, the response of the chromatographic detector should always be considered. Hexane or petroleum ether will appear as the predominant peak in the subsequent chromatogram if a gas chromatograph with flame ionization detection is used. The least interference will be caused if the solvent peak appears before the analyte peaks but there is still a potential problem with peaks resulting from trace impurities in the solvent. Because of this, even analytical-grade solvents may have to be redistilled prior to use.

If a selective chromatographic detector is being used, it is possible to use an extraction solvent for which the detector has low sensitivity, e.g. hexane or petroleum ether for electron capture detection with gas chromatography. Aromatic solvents should be avoided if liquid chromatography with ultraviolet detection is to be used.

If an unsuitable solvent cannot be avoided (e.g. if a chlorinated or oxygenated solvent is required with subsequent electron capture detection) and the analyte has low volatility, it is possible to evaporate the extract to dryness and redissolve the residue in a compatible solvent. It is, however, better to avoid this if at all possible.

Liquid–liquid extraction has long been seen as the standard extraction method but in more recently developed procedures it is being replaced by newer techniques such as solid-phase extraction (SPE). These are more rapid and can more easily be automated.

4.2.2.2 Solid-phase Extraction

The use of this technique has rapidly increased over the past few years and when developing new procedures may often be the first-choice method. A short disposable column containing 100–500 mg of adsorbent material is used here. The column packing is usually a reversed-phase material similar to that used in high performance liquid chromatography columns. The use of an ODS packing material (which contains octadecylsilane groups chemically bonded on to a silica support) is common. Other materials are available, including ion exchangers and adsorbents such as 'Florisil'.

Before use, conditioning of the column is usually necessary – this is carried out by passing a small volume of methanol through the column. Preconditioned columns are, however, commercially available. The water sample is then passed through the column by applying mild suction or pressure. The organic components of the sample are retained on the packing material. The column can then be washed with water or another suitable solvent to remove potentially interfering compounds, and air dried if the wash solvent is immiscible with the following solvent. The compounds of interest are then eluted with a few millilitres of a suitable organic solvent. As sample volumes could be several hundred millilitres, a concentration factor of about 100 is routine (Figure 4.2).

A wide range of solvents can be used. The best extractants are often those where the polarity of the solvent matches that of the extractant, e.g. hexane could be used for non-polar organochlorine pesticides. The subsequent stage of the analytical procedure could also influence the choice of solvent. Methanol or acetonitrile are often used if liquid chromatography is to be used as the separation method. The procedure for solid-phase extraction is summarized in Table 4.1.

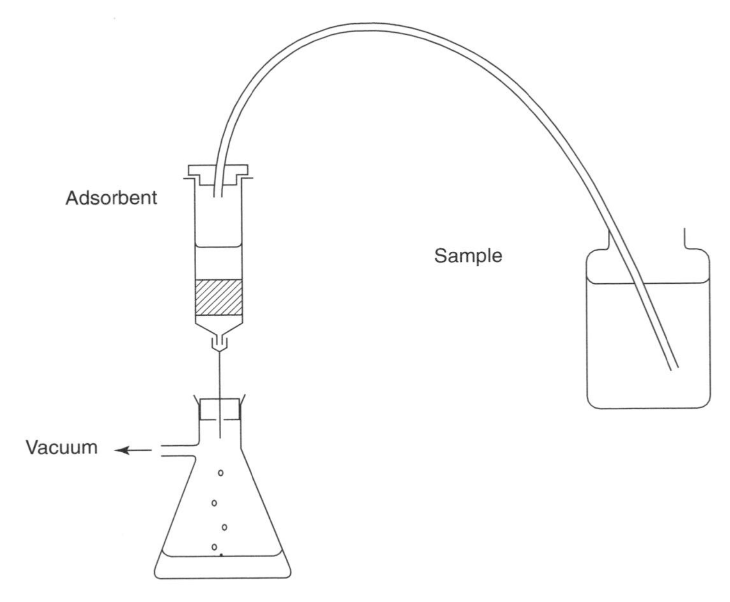

Figure 4.2 Solid-phase extraction with large sample volumes.

Table 4.1 Steps in a solid-phase extraction process

- Condition column with methanol
- Load sample
- Wash column with water
- Pass air through the column to remove as much water as possible (an option if the elution solvent is immiscible with water)
- Elute with a suitable organic solvent

Manifolds are available which allow processing of a number of samples simultaneously. In addition, SPE set-ups can be directly coupled to HPLC systems.

DQ 4.5

What do you consider the advantages of solid-phase extraction over liquid–liquid extraction which has now made this the first-choice method?

Answer

- *It is a very rapid process and can easily be automated*
- *High concentration factors can easily be achieved*
- *Solvent consumption can be much lower*

There are, however, instances where liquid–liquid extraction may still be the method of choice. These are usually when there are solids present or there is a high loading of organic material in the sample (e.g. humic acid) which could block or overload the column.

A further development is the use of extraction discs where the adsorbent material is held within the fibre structure of a polytetrafluoroethylene (PTFE) filter disc. After pre-washing the disc with a portion of the final eluting solvent and conditioning with methanol, the extraction procedure is simply to pass the sample, by suction, through the filter. The extracted components are then eluted by using a suitable solvent. The advantages of discs over columns include a higher sample throughput (several hundred millilitres of sample may need to be passed through the filter if a high concentration factor is needed) and the lower likelihood of the filter clogging with particles. Some standard methods now include liquid–liquid extraction and solid-phase extraction as alternative procedures.

4.2.2.3 Head-space Analysis

In this technique, the water sample is placed in a container with a septum seal in the lid and an air space above the sample. The most simple procedure is then, after allowing for the air to equilibrate with the water, to inject an air sample (containing volatile organic components) into the gas chromatograph. This technique overcomes problems found in liquid–liquid extraction resulting from solvent interference. The sensitivity towards a particular component will, however, be dependent on its volatility, favouring low-molecular-mass, neutral components. The overall sensitivity of the technique may be increased by heating the sample. Be aware, however, that you are also increasing the vapour pressure of the water and care should be taken to check the water compatibility of the chromatographic column.

4.2.2.4 Purge and Trap Techniques

These techniques extract the volatile organic content from the sample by using a purge gas stream. In many instruments, the organics are collected in a short tube of adsorbent material such as activated charcoal or a porous polymer (e.g. 'Tenax'). After the collection period, the tube is flash-heated to release the organics into the gas chromatograph. Other instruments collect the volatile components into a secondary liquid nitrogen cold trap. Rapid heating of this trap then releases the organics into the chromatograph.

4.2.2.5 Solid-phase Microextraction

This technique could be seen as using both the principles of solid-phase extraction and head-space sampling. A fibre which is originally contained within a syringe needle (Figure 4.3) is exposed either to the stirred sample or to the head-space above the sample. The fibre typically consists of fused silica with a coating of polydimethylsiloxane, or alternatively polyacrylate, with the phase being chosen according to the compound being determined. The dissolved components partition between the sample and fibre. After equilibration is complete (2–15 min for liquid samples), the fibre is withdrawn into the syringe needle for storage prior to analysis.

The method cannot only be used for volatile organics but also for semi-volatile pollutants, such as chlorinated pesticides. Different fibres are used. A smaller-depth coating (7 μm) is suitable for the semi-volatile compounds, and a thicker (100 μm) coating for volatiles. Fibres are also available for the extraction of polar organic compounds (e.g. phenols) which are often very difficult to extract by other techniques. In the case of phenols, sample modification (e.g. lowering the pH and adding sodium chloride) increases the extraction efficiency.

Subsequent analysis can be carried out by either gas or liquid chromatography. With gas chromatography (GC), the fibre is directly introduced into the GC injector inside the syringe and re-exposed once the needle has pierced the injection septum. Most GC systems can be used without modification. Desorption of the organics takes place into the carrier gas, although this can take 20–30 s. In order to overcome this problem, a technique known as *cryo-focusing* is used. The sample is condensed on to the top of the column held at a low temperature, typically 40°C. Rapid heating of the column then releases the sample. At least one manufacturer offers a solid-phase microextraction desorption apparatus integrated with a GC system.

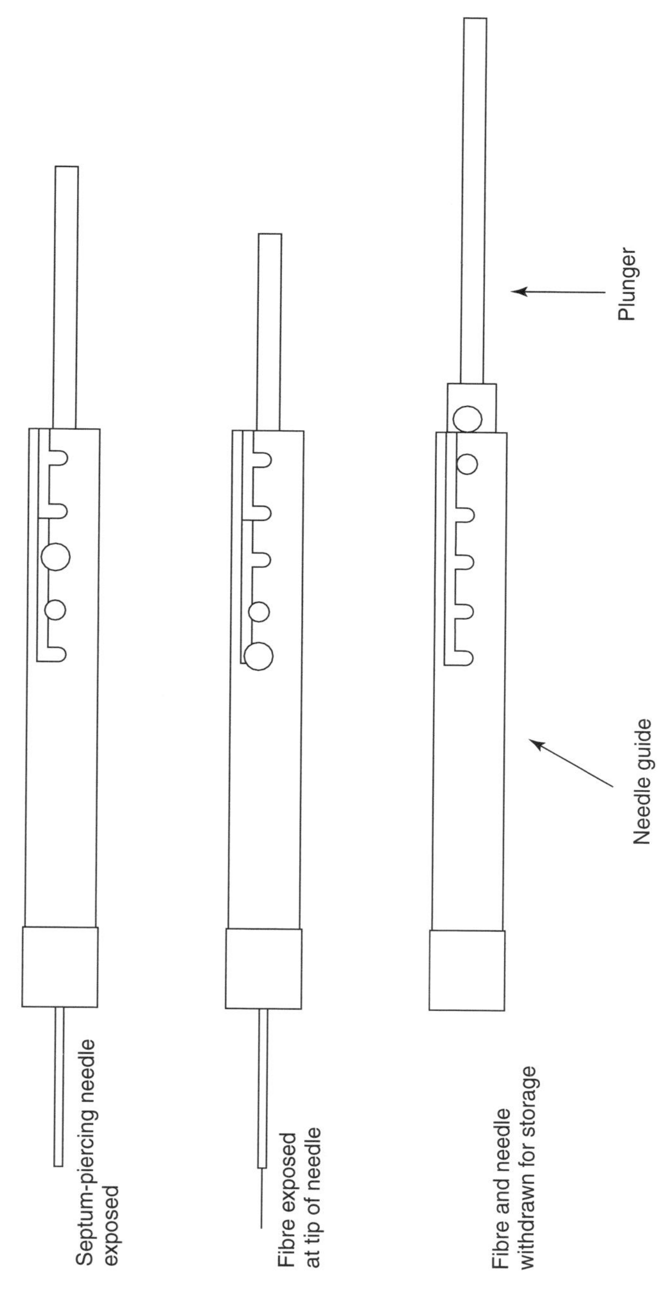

Figure 4.3 Schematic of the solid-phase microextraction process.

If the subsequent chromatographic method is high performance liquid chromatography (HPLC), then the compounds can be desorbed by immersion of the fibre into a suitable solvent. The solution is then injected into the chromatograph. Injection systems are also available which permit the introduction of the fibre directly into the mobile phase, where the latter flows along the length of the fibre on to the head of the analytical column.

The fibres can be re-used as many as 50–100 times. The advantages of the technique include its simplicity and low cost of apparatus. As the complete extract is introduced into the chromatograph, this can lead to 100–700× lower detection limits than liquid–liquid extraction. No solvent is injected and short narrow-bore columns can be used with gas chromatography (see Section 4.2.3 above). These columns would become flooded with solvent if used after liquid–liquid extraction. The fibres can cope with high levels of contamination and so they can be used for dirty samples such as waste water. One disadvantage of the method is that it is an equilibration technique. Extraction of each compound will be different and so calibration is necessary for each of these. In addition, changes in composition of the water samples could alter the extraction equilibria and hence the extraction efficiency.

DQ 4.6

What are the advantages and disadvantages of immersing the fibre into the sample and sampling the head-space?

Answer

Head-space sampling would give a cleaner extract for volatile and semi-volatile samples. Direct immersion is a simpler method but it could suffer from blockages if there are suspended solids in the sample.

4.2.3 Gas Chromatography

DQ 4.7

From your knowledge of this technique, sketch the major components of a typical gas chromatograph. What is the principle by which the separation occurs?

Answer

Chromatographic separation of a mixture occurs by the differential partition of the components between a stationary phase and a mobile phase. In gas–liquid chromatography, the mobile phase is a gas and the stationary phase is a liquid adsorbed on, or chemically bonded to a solid. The main components of a gas chromatograph are shown in Figure 4.4.

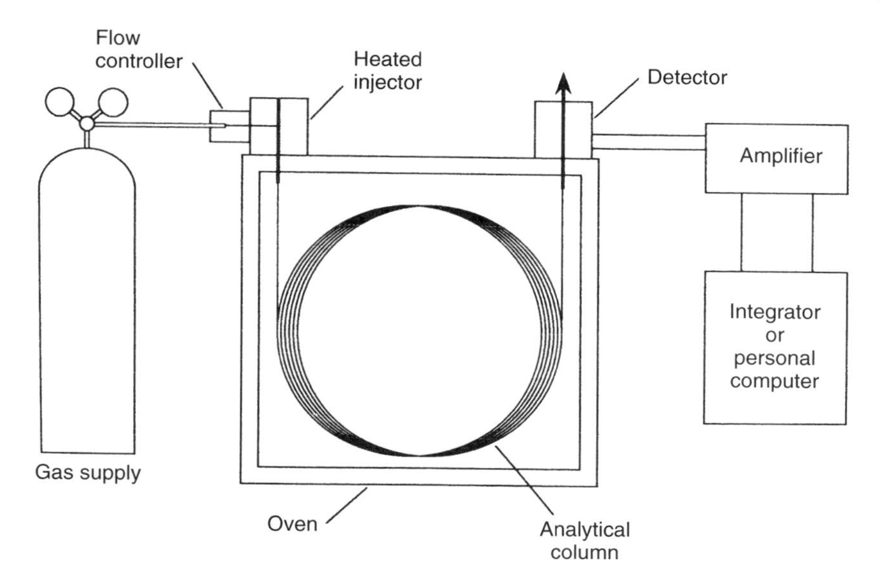

Figure 4.4 Major components of a gas chromatograph.

Gas chromatography has the advantage over other chromatographic techniques of combining high separation efficiencies with the availability of highly specific and sensitive detectors. A high proportion of the separations required can be performed by using just a few stationary phases. The wide range of phases which are available does, however, permit the development of columns for specific problem separations.

We will discuss the columns and detectors used for water analysis first of all, and then present some examples of complete analytical procedures (including sample pretreatment).

4.2.3.1 Detectors

The most common detectors used for environmental trace analysis are listed in Table 4.2.

The electron capture detector has held a special place within environmental analysis since many of the compounds of concern contain chlorine atoms. Had it not been for the development of this highly sensitive and specific detector (for some compounds, 10–100 times more sensitive than flame ionization detection) much of the trace analysis required for these compounds would not have been possible.

More recently, mass spectrometry has found use as a sensitive and highly selective detection method. The detector can produce a chromatogram which is selective to a particular mass (or more accurately mass/charge ratio),

Table 4.2 Common detectors used in gas chromatography

Detector	Typical Application
Flame ionization detector	Sensitive universal detector for organic compounds
Electron capture detector	Highly sensitive, specific detector responding to atoms with a high electron affinity e.g. chlorine. Typical analytes are chlorinated pesticides and chlorinated solvents
Hall electrolytic conductivity detector	Highly sensitive, specific detector for halogens, nitrogen and sulfur. Typical analytes are pesticides and trihalomethanes
Thermionic detector	Element-specific detector for compounds containing nitrogen and phosphorus. Typical analytes include pesticides
Flame photometric detector	Element-specific detector for compounds containing sulfur and phosphorus. Typical analytes include pesticides
Photo-ionization detector	Specific to compounds with aromatic rings or double bonds. Typical analytes include industrial solvents
Mass spectrometric detector	Highly specific, sensitive detector for all organic compounds. Can also be used for peak identification

thus simplifying the chromatogram greatly and to some extent lessening the requirements for pretreatment. Although the potential of this technique was apparent for many years, its widespread application had to await the development of low-cost bench-top gas chromatograph/mass spectrometer (GC–MS) systems (using quadrupole or ion-trap spectrometers) rather than the more expensive and cumbersome combination of separate instruments. Advances had also to be made in the availability of cheap computer data processing and storage facilities to handle the massive amount of information produced from even a single chromatographic separation. This method is becoming increasingly routine and, in many laboratories, GC–MS is now the standard technique. Simple applications are described here, particularly in how GC–MS can aid quantification, although a more detailed discussion of the technique is left to Chapter 8 (Ultra-Trace Analysis), where its advantages in such applications are of the greatest importance.

4.2.3.2 Columns and Stationary Phases

The range of columns available is extensive. In the choice of column for a particular application, not only does the chromatographer have to consider the most appropriate stationary phase but also the column dimensions. The latter not only affects the separation efficiency, but must also be considered for compatibility with the detector being used, the method of sample introduction, and the sample type.

The column types available can be divided into the following, listed in order of decreasing separation efficiency:

- *Narrow-bore Capillary Columns.*
  Typical dimensions: length, 30–60 m; i.d., 0.2 mm; flow, 0.4 ml min^{-1} He
- *Wide-bore Capillary Columns.*
  Typical dimensions: length, 15–30 m; i.d., 0.53 mm; flow, 2.5 ml min^{-1} He
- *Packed Columns.*
  Typical dimensions: length, 2 m; i.d., 2 mm; flow, 20 ml min^{-1} He

Most recent analytical methods for water analysis use the first two types of column, but you may occasionally find packed columns in long-established methods or for less-demanding applications.

Narrow-bore columns offer the greatest detection sensitivity and are used for analyses close to the limits of detection. The low carrier gas-flow rate is well suited for applications where mass spectrometric detection is used. However, direct sample injection on to the column by using a syringe is not possible as the column would become overloaded. A splitting device is necessary for the introduction of the sample.

Wide-bore columns have a larger sample capacity and direct syringe injection is possible. The sample may also be introduced from a sample concentration system such as a 'purge-and-trap' device. The greater sample capacity may be required if a low-sensitivity detector is being used. Wide-bore columns are also less affected by contamination from non-volatile components in the sample and so find a use with highly contaminated samples, such as waste water.

Many of the organic compounds of environmental interest are of high relative molecular mass and have low volatilities. High oven temperatures are necessary for these and consequently silicone polymers are often the favoured stationary phases. Poly(ethylene glycol) columns are also popular. As with other uses of gas chromatography, the best separation efficiencies are achieved when the stationary phase has a similar polarity to the components of the analyte. Fuel oils are separated on non-polar columns (e.g. dimethylsilicone), pesticides and chlorinated solvents are often separated on medium-polarity columns (e.g. diphenyl/dimethyl silicone), whereas 2,3,7,8-tetrachlorodibenzo-*p*-dioxin can be separated from its isomers by using highly polar columns (e.g. cyanopropyl silicone).

The stationary phase may be adsorbed or chemically bonded on to the column walls of capillary and wide-bore columns, or on to a support material in packed columns. For analyses close to the limit of detection and at high oven temperatures, column bleeding may become a significant factor. The use of low-loaded columns (0.1–0.25 μm film thickness), or chemically bonded phases may reduce this effect. A higher loading of columns (1–5 μm film thickness) is possible at lower temperatures for the analysis of volatile compounds. Thicker films have higher sample capacities for highly concentrated components, but there is a corresponding decrease in column efficiency when compared to thinner films.

DQ 4.8

How would you confirm that a peak is due to a single component rather than two components with identical retention times?

Answer

A chromatogram should be produced on two columns of different polarities. It would be unlikely that the peaks would remain unresolved on both columns.

Many standard procedures specify the use of two columns, with the second column being known as the *confirmational column*. It is for this type of problem that capillary columns show their greatest advantage over packed columns. Their greater separation efficiency reduces the probability of unresolved peaks.

4.2.3.3 Injection Methods

If you are using a narrow-bore capillary, then a device is needed to reduce the microlitre volumes injected by syringe to the nanolitre volumes which the column can accept without overloading. A number of techniques are available, including those summarized below. The first two methods use a split/spilt-less injector (an injection system which can be used in either of the two modes), whereas the third requires a modified form which has simple temperature programming.

With *split injection*, the sample from the syringe is introduced into a vaporizing chamber which is maintained at a high temperature and has a lateral through-flow of gas. Only a small fraction of the sample enters the column, with the rest escaping to the atmosphere through an outlet valve.

With *split-less injection*, the full sample is vaporized before introduction into the column, which is held at a temperature below the boiling point of the solvent. This concentrates ('focuses') the sample in a small section of the capillary so that when the temperature programme is begun the solvent elutes as a narrow band without interfering with the analyte peaks. Apart from venting the gases at the end of the transfer on to the column, the whole sample is transferred. This makes the technique more sensitive than the split method.

The final technique, i.e. *large-volume injection*, is useful for trace analysis, as a concentration stage is included. Up to 250 μl of sample is slowly injected on to a cold short column which may be packed by capillaries or packing material. Most or all of the solvent is slowly vaporized (20–30 s) before a more rapid heating to transfer the concentrated sample on to the column. The latter is held at a low enough temperature to focus the sample in the capillary. The chromatographic separation starts on commencement of the temperature programme.

4.2.3.4 Examples of Analytical Procedures

Most analytical methods involve the extraction of the compounds from water before chemical analysis. We have already noted two reasons for this – separation of potential interferences and use as a concentration stage.

DQ 4.9

Can you think of two further reasons which are specific to gas chromatography?

Answer

Many, but by no means all, gas chromatographic columns and detectors are incompatible with water.

Direct injection of the sample would deposit non-volatile solids on the column, which could cause blockage and would shorten column life.

With a suitable choice of analytical column, simple extraction may be a sufficient pretreatment for the direct injection of the extract into the chromatograph. For instance, the UK HMSO method for halomethanes simply uses the extraction of the compounds into petroleum ether and injection of the extract directly into the chromatograph. The chromatograms of such extractions may be complex (particularly if flame ionization detection is being used), with a single extraction stage usually having insufficient specificity to simplify the chromatograms greatly. Indeed, a simple extraction with injection of the extract into the chromatogram is often used as a survey method to identify organic compounds in water.

DQ 4.10

What alternative extraction methods would be suitable for halomethanes?

Answer

Halomethanes are examples of volatile organics. Head-space analysis, purge and trap or solid-phase microextraction would be suitable.

For the analysis of individual components (often semi-volatiles) expected to be found at low concentrations (e.g. pesticides), further pretreatment may be necessary.

DQ 4.11

What are the major stages in any pretreatment scheme?

Answer

1. *Extraction*
2. *Clean-up to remove interfering components*
3. *Concentration of extract*

(see Section 2.7 above)

Until the successful development of solid-phase extraction, solvent extraction had been the most often used technique for stage one. The low volatility of many of the compounds of interest in this category renders the alternative vapour-phase extraction methods difficult. The clean-up stages will invariably be chromatographic, often using column chromatography. This may involve more than one separation stage. To illustrate the method, it is easiest to study one analysis in detail. For this, I have chosen an analytical scheme for the commercial pesticide, DDT. This is taken from the European Standard Method, EN ISO 6468 (1996).

Analysis of DDT. This was the first synthetic insecticide to come into widespread use. It was introduced after the Second World War, and although now controlled or banned in many areas of the world (particularly in the West), it is now a universal contaminant. In common with most commercial products, the insecticide is not a single chemical compound – the major active component (*p*,*p*′-DDT) only consisting of 70–80% of the total content. One of the minor components, *p*,*p*′-DDD (similar in structure to *p*,*p*′-DDT, but with a $-CHCl_2$ side-chain rather than $-CCl_3$) is, in fact, more toxic to insects than *p*,*p*′-DDT.

When considering environmental samples, a number of decomposition and metabolic products will also be present. Some of the reactions producing these materials have been considered above in Section 2.3. In fact, for many samples, the highest-concentration component is not *p*,*p*′-DDT but its primary metabolic product, DDE.

The following chromatographic peaks are expected in DDT analysis:

(i) Components of technical DDT

p, *p*′-DDT (70–80%)

o, *p*′-DDT (15–20%)

p, *p*′-DDD (1–4%)

(ii) Decomposition products

p, *p*′-DDE (aerobic decomposition)

p, *p*′-DDD (anaerobic decomposition)

Thus, we have a multi-component mixture even without the presence of any other compounds expected in the water sample! Interfering components in a typical sample could include other pesticides and polychlorinated biphenyls (PCBs). These often have similar extraction properties to the DDT components.

A typical pretreatment would be as follows:

1. Extraction of the organic components into hexane, with a 1 l sample being extracted into three aliquots (30 + 20 + 20 ml) of solvent.

2. Drying the combined extracts by using a column containing 5 g sodium sulfate. The chromatographic columns in the subsequent stages of the procedure are deactivated by the presence of water and so drying the extract at the earliest possible stage is essential.
3. Further concentration of the extract to a 1 ml volume. This could be by a number of methods, including using a Kuderna–Danish evaporator or a rotary evaporator. The sample is then placed on to the top of the first chromatographic column described below.
4. Clean-up of the extract by column chromatography.

   (a) *Alumina–alumina/silver nitrate column.* This column contains a bottom layer of alumina/silver nitrate, a layer of alumina and a top layer of sodium sulfate. Alumina is a polar column material and will retain polar components in the extract. The silver nitrate helps retain compounds containing unsaturated carbon–carbon bonds. Non-polar material, including the DDT components, is eluted by using 30 ml of hexane. The extract is next reduced in volume to 1 ml and then a 100 μl portion is added to the top of the second column.

   (b) *Silica gel column.* This is a less polar column than the first and can be used to separate potential non-polar interferences from the sample. First, 10 ml of hexane are passed through the column, eluting the PCBs, with the DDT components being retained on the column, followed by 8 ml of a 90% hexane/10% toluene solvent mixture. DDT is then eluted with a more polar solvent mixture (12 ml of 10% diethyl ether in hexane).

The eluates are then re-concentrated to 1 ml or less before injection into the chromatograph. Try calculating the overall concentration factor of the pretreatment process. You should come up with the factor of 100× if the final solution volume is 1 ml. A typical chromatogram is shown in Figure 4.5. The detection limit for each component is approximately 10 ng l^{-1}.

The above procedure is just one method of pretreatment. Other chromatographic methods may be used, such as preparative-scale thin layer chromatography (TLC) or solid-phase extraction (SPE). Each of these methods will still, however, be made up of the same individual stages of extraction, concentration and removal of selected interfering components.

Later chapters will extend the use of this method to the analysis of solids (Chapter 5) and to ultra-trace components (Chapter 8).

DQ 4.12

Clean-up of the extract simplifies the subsequent chromatogram. Can you think of a second advantage?

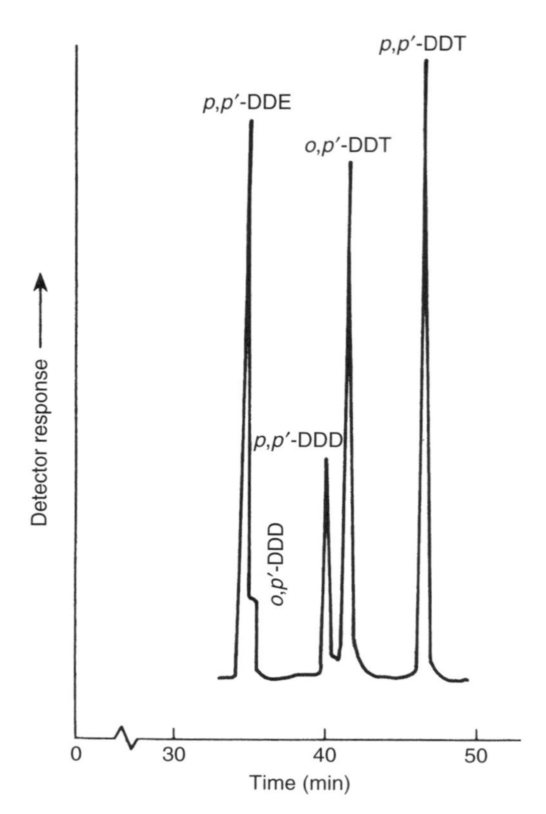

Figure 4.5 Chromatographic separation of DDT components using a 25 m × 0.32 mm i.d. methylsilicone capillary column with a temperature gradient to 220°C.

Answer

This is simply protection of the column and detector from contamination. Without clean-up, the column lifetime will be shortened and the detector sensitivity lowered. Cleaning detectors to restore the sensitivity can be very time-consuming!

Fingerprinting Oil Spills. If a film of oil is discovered on water, the first question likely to be asked is:

'What is it? Petrol? Fuel oil? Paraffin?'

The next question might be:

'Where did it come from?'

The commercial products mentioned above are complex mixtures of organic compounds. The precise composition of the mixture can vary from sample to sample, and so even if a complete quantitative analysis of every component in the mixture were undertaken, there would still be much difficulty in interpretation of the data.

A simpler procedure is to produce a chromatogram under standard conditions (column packing, flow rate, column temperature, etc.) and to compare the trace either with a library of reference materials, or preferably a sample of the material suspected to have been discharged. Capillary columns are necessary for resolution of individual components. Often, the correspondence of retention times and the overall envelope shape of the chromatogram will be sufficient to characterize the effluent. Hydrocarbon fuels give chromatograms with regularly spaced peaks (consecutive members of homologous series of compounds within the fuel). Lubricating oils have fewer resolved peaks. Natural product (vegetable) oils have simpler chromatograms with few individual peaks. Further information can be obtained if individual components in the material can be identified. Thus, the presence of simple polyaromatic species, such as anthracene, will identify coke-oven fractions.

Sample preparation from water containing low concentrations of hydrocarbons is simply to extract the material with a suitable volatile solvent (e.g. diethyl ether). After washing and drying, the extract is concentrated by using a dry nitrogen stream. With heavily polluted water, the organic material is separated by extractive distillation with toluene. The oil can then be recovered by fractional distillation from the toluene.

Complications can occur in interpretation of the chromatogram with material which has not been sampled immediately after discharge. The compositions of oil spills change with time (see Figure 4.6). Volatile components of the oil evaporate, with, in general, lower-molecular-mass components disappearing first. The oil will also slowly be biodegraded, with the rate of degradation of a particular component being dependent on its chemical structure. A straight-chain hydrocarbon, for example, will be degraded more quickly than its branched-chain isomer. The determination of trace components and their relative concentrations has been found useful for assisting identification. Polynuclear aromatic hydrocarbons and their alkylated derivatives have been used as reference compounds in the finger-printing of oil spills. Although GC apparently seems a very simple method of identification of spillages, it is, in fact, a skilled task needing a great deal of experience.

4.2.3.5 Quantification

When you look at a number of gas chromatography standard methods it almost seems that each has a different procedure to determine concentrations. Most are variations of one or more of the methods described below. We will start with the most simple method, discuss the problems with this approach, and then consider some ways of overcoming the problems.

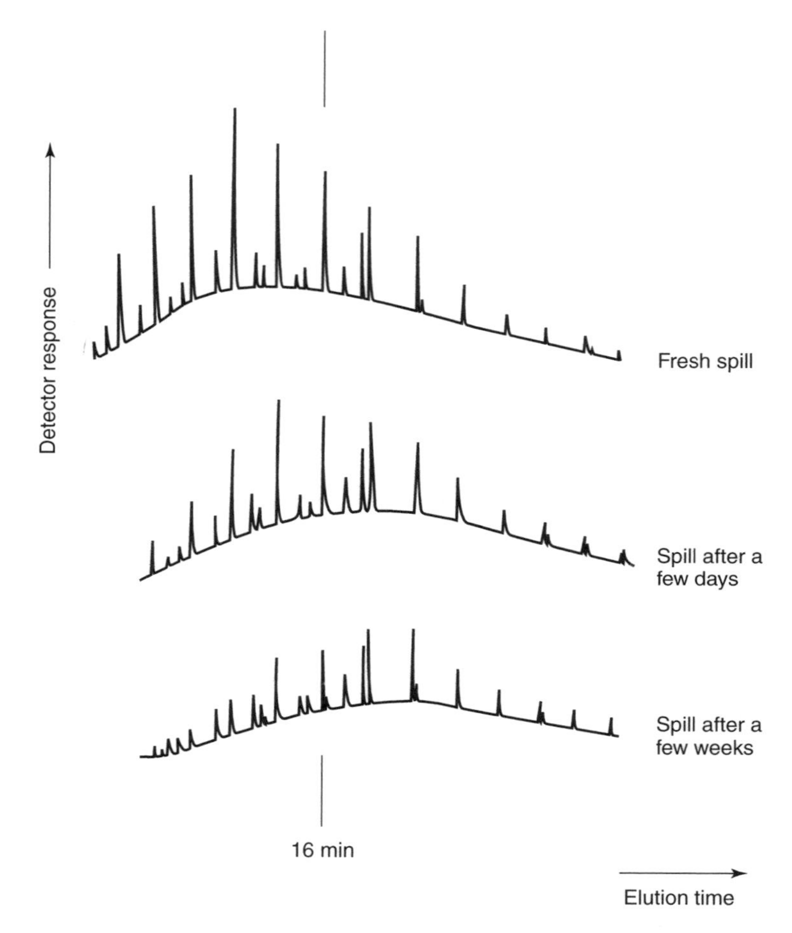

Figure 4.6 Typical envelope shape in a chromatogram of an oil spill and its change with age of spill.

External standards. The simplest and most obvious method is to compare the peak area of the compound in the unknown solution with the areas of a series of solutions which have been used to form a calibration curve (or, if the calibration has been shown to be linear, with a single solution of known concentration close in value to the unknown), i.e. calibration is by external standards as discussed earlier in Section 3.4.1.1.

DQ 4.13

After consideration of the ways a sample can be introduced into a gas chromatograph, why do you think the method of external standards may not be ideal?

Answer

The external standard method assumes that the same volume of solution is being introduced during each injection. It is extremely difficult to introduce reproducible sample volumes into the chromatograph.

Internal standards. The above problem can be overcome by adding an *internal* standard. This is a compound which will produce a chromatographic peak close to, but resolved from, the unknown species. An accurately known amount of the standard is added to a fixed volume of the unknown solution and to each of the external calibration solutions. Any variation in injection volume would show up in a change in peak area of the internal standard. There are a number of methods by which a correction may be applied. The normalization procedure plots the following:

$$\frac{\text{peak area}}{\text{internal standard peak area}} \text{ versus concentration}$$

DQ 4.14

Plot a calibration graph from the following data and determine the concentration of the unknown.

Concentration(μg l^{-1})	10	20	30	40	50	Unknown
Peak area[a]	2315	3800	5900	8680	11600	5570
Internal standard peak area[a]	6150	5900	5740	6150	6730	6050

[a]In arbitrary units.

Answer

The normalized peak areas are as follows:

Concentration ($\mu g\ l^{-1}$)	*10*	*20*	*30*	*40*	*50*	*Unknown*
Area	*0.3764*	*0.6441*	*1.028*	*1.411*	*1.724*	*0.9207*

The least-squares line is given by:
Peak area ratio = 0.034 62 × (concentration) – 0.001 93
This gives the concentration of the unknown as 26.5 $\mu g\ l^{-1}$.

Quantification if there is sample pretreatment. There is the potential for loss of analyte during the pretreatment. The procedures described so far will not take this into account and any loss of the unknown will result in a low analytical value. One method to overcome this problem would be to use external standards and to submit them to the same clean-up procedure as the unknown. Losses during clean-up would be assumed to be the same for the standards and the unknown. Internal standards would again be added prior to chromatography to overcome injection problems.

During method development and as a quality control step, the percentage recovery would need to be determined. The procedure for this is described below. Such a value should be as close to 100% as possible, although for some complex extractions, values of more than 60% may be acceptable. You could have little confidence in the result if you discovered that 90% of the unknown was being lost during the clean-up!

Percentage recovery. A known amount of the compound being determined is added to a blank. The latter could be a synthetic solution made up as close as possible to the expected sample composition (not including the unknown), or it could be a field sample from which the analyte has been extracted (a 'pre-extracted' sample). After extraction and clean-up, the sample is injected into the chromatograph and the peak area (or more precisely, the normalized peak area, as internal standards are necessary) produced is compared to that expected from a directly injected compound. You should remember to take into account any sample concentration during pretreatment. The recovery is determined from the following:

$$\frac{\text{peak area found} \times 100}{\text{peak area expected}}$$

Isotope dilution analysis. Unless the standard solutions were exactly matched in composition to the unknown, there would still be the possibility of errors due to the sample matrix changing the recovery efficiency. In order to overcome this, you would need to have the standards and the unknown in the same matrix, i.e. you need to have the *standard solutions added to the sample*. This is possible if mass spectrometric detection is used. The standard is an identical compound to the unknown except that one or more atoms have been isotopically substituted – this is often a deuterated compound. The peaks corresponding to the unknown and the standard will have the same retention times, but can be distinguished by detection at the two mass/charge ratios corresponding to the unsubstituted compound and the isotopically substituted standard. This is illustrated in Figure 4.7.

In practice, allowances have to be made in the calculation for the compound and the substituted compound not being isotopically pure, i.e. detection at any one mass would include contributions from both the sample and the 'spike' (see Figure 4.8). The isotopic abundances of the unspiked sample and the spike need to be precisely known from the mass spectra of the individual components (scans (a) and (b) in Figure 4.8.). The concentration in the original sample can then be calculated from equations derived from the ratios given in the following

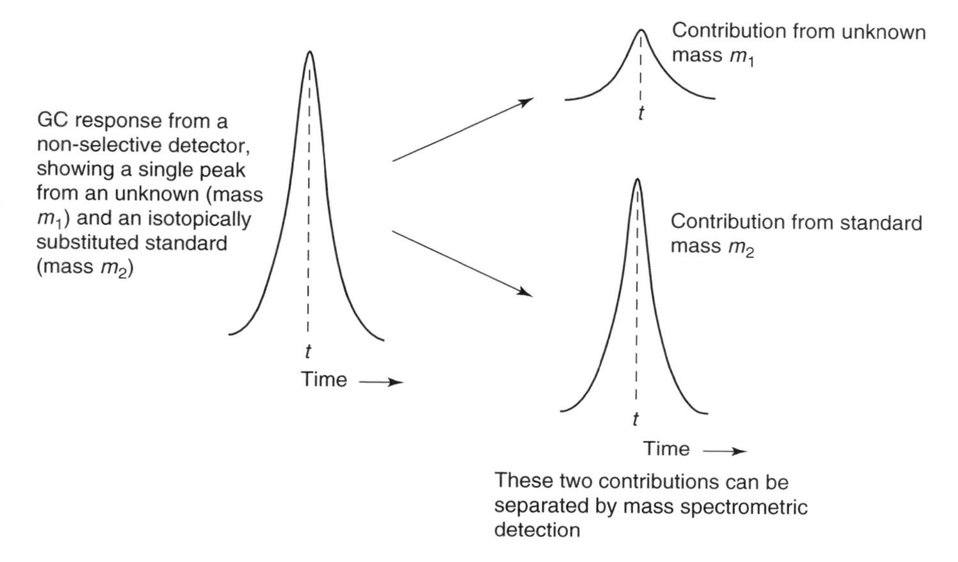

Figure 4.7 Illustration of the principle of isotope dilution analysis.

equation:

$$\text{Intensity ratio (mass } m_1/\text{mass } m_2\text{) of spiked sample}$$
$$= \frac{\text{number of unknown molecules at mass } m_1 + \text{number of spiked molecules at mass } m_1}{\text{number of unknown molecules at mass } m_2 + \text{number of spiked molecules at mass } m_2} \quad (4.3)$$

where the number of unknown (or spiked) molecules at mass m_x ($x = 1$ or 2) is given by:

$$\frac{\text{total mass of unknown (or spike)} \times \text{fractional molecular abundance at mass } m_x}{\text{molar mass of compound (or spike)}} \quad (4.4)$$

You should note that although isotope dilution analysis overcomes many of the problems found in other methods there should ideally be one isotopic standard for each compound being determined. However, this may increase the cost of the analysis quite considerably.

4.2.4 Liquid Chromatography

For several years, liquid chromatography (LC) had the role of separation of classes of compounds which was difficult to achieve by using gas chromatography. Its application is now widening, largely due to the advent of bench-top

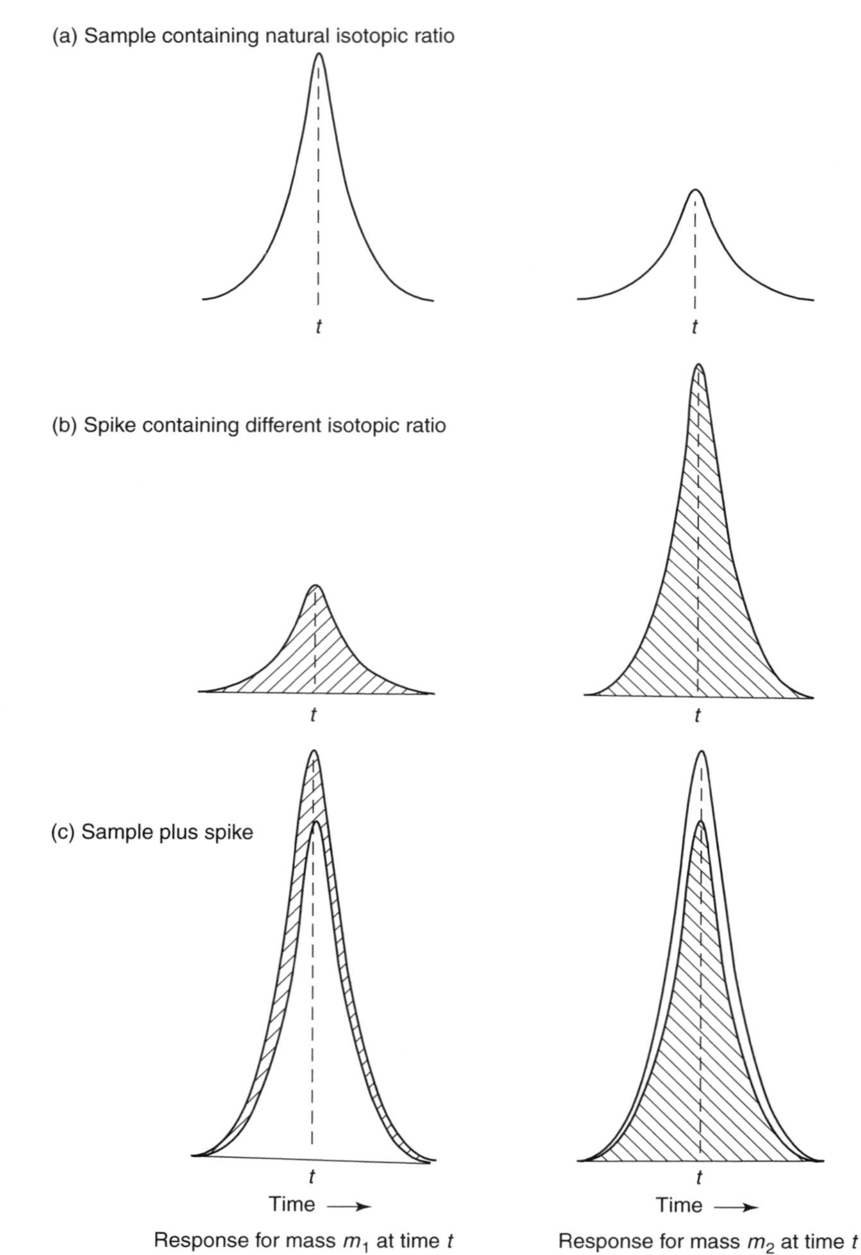

Figure 4.8 Quantification by isotope dilution analysis in practice.

LC–MS systems and the increasing use of solid-phase extraction which is particularly suited to interfacing with LC equipment. The lower separation efficiency of LC in comparison to that of GC is largely offset by the high selectivity of mass spectrometric detection.

Longer established methods which use LC concern groups of compounds which can be determined using specific detectors or by derivatization.

DQ 4.15

The most common form of high performance liquid chromatography uses ultraviolet absorption as its method of detection. From your knowledge of liquid chromatography, what alternative detection techniques may find use in environmental analysis?

Answer

Conductivity detection can be used for ionic species (see Section 3.4.3 above). Fluorescence detection has an extremely high sensitivity and selectivity to specific groups of compounds and may find use for such species.

Conductivity detection has found widespread use for inorganic ions. Low-molecular-mass carboxylic acids (e.g. formic and acetic acids) have very similar physical properties to the inorganic acids and ion chromatography provides a convenient alternative to gas chromatography for these acids.

One group for which fluorescence detection has high sensitivity are polynuclear aromatic hydrocarbons (PAHs). Some examples of these are shown in Figure 4.9. They are highly carcinogenic compounds which are produced in trace quantities whenever fossil fuels are burnt. Typical water extracts could include up to 70 PAHs with a total concentration of around 1 $\mu g\ l^{-1}$. In order to monitor these low concentrations, sample preconcentration is needed. Solid-phase extraction, using an octadecylsilane (ODS) column, or a combination of ODS and amino-type columns, has been used. Sensitivity can be maximized if the detector is capable of changing the excitation and detection wavelengths throughout the chromatographic run, since each component has different optimum settings. The range of wavelengths used is 270–300 nm for excitation and 330–500 nm for detection.

Fluorescent derivatives can be made from non-fluorescent or weakly fluorescent compounds. Phenols and *N*-methylcarbamate pesticides (Figure 4.10) are often analysed in this way. The procedure for *N*-methylcarbamates uses *post-column derivatization*. The HPLC eluent is hydrolysed with sodium hydroxide at 95°C, thus producing methylamine. The latter is then reacted with *o*-phthalaldehyde and 2-mercaptoethanol to produce the fluorescent derivative. The fluorescent excitation wavelength is 230 nm and detection is $>$ 418 nm, giving a limit of detection of approximately 1 $\mu l\ g^{-1}$ per component for a 400 μl sample, injected without preconcentration.

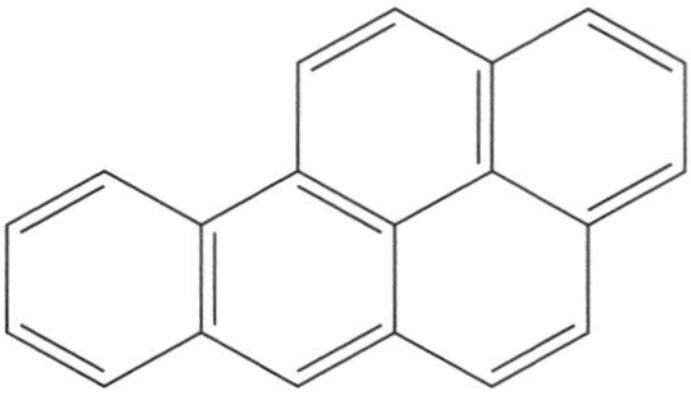

Benzo[*a*]pyrene

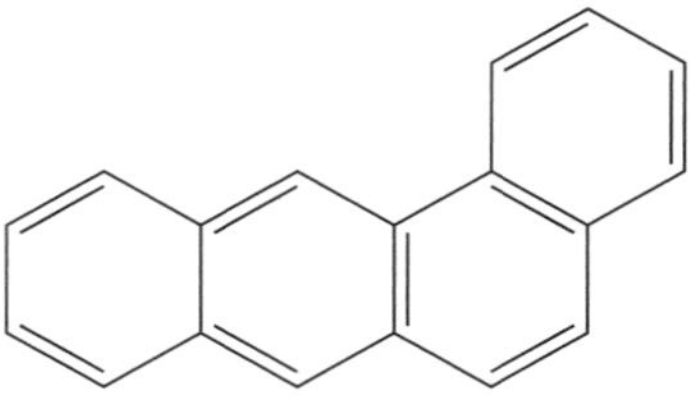

Benz[*a*]anthracene

Figure 4.9 Some typical PAHs found in environmental samples.

Type	*Example*	*Structure*
N-Methylcarbamate	Carbofuran	
Urea	Diuron	
Triazine	Atrazine	

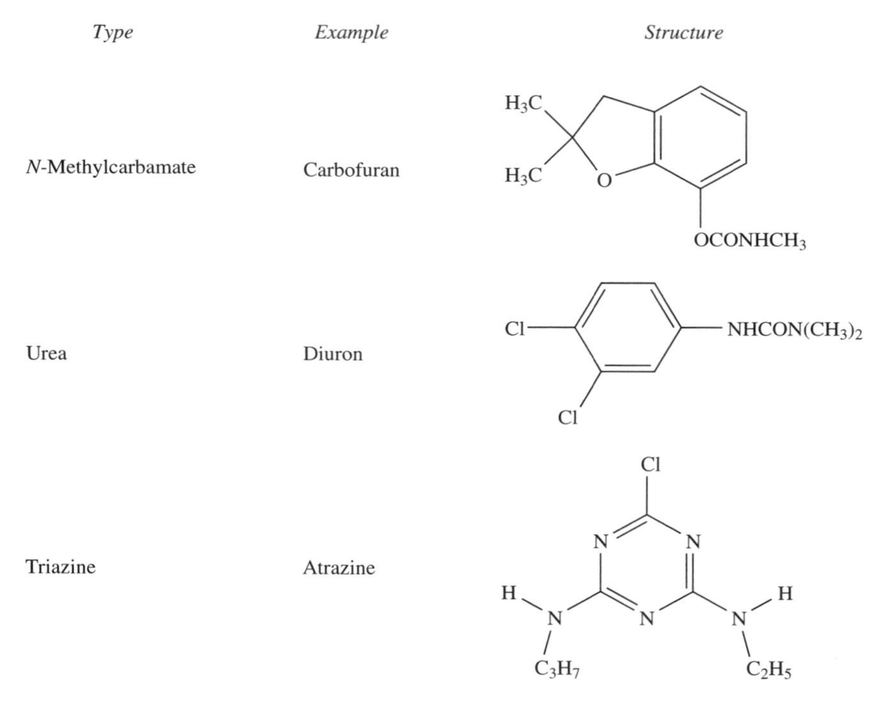

Figure 4.10 Some examples of pesticides that can be analysed by using liquid chromatography.

HPLC with ultraviolet detection is sometimes used for these and similar species, e.g. *N*-methylcarbamate, urea and triazine pesticides can be analysed by this method. These are 'second-generation' pesticides which have been developed to replace organic halogen compounds. The sensitivity with UV detection is lower than that achieved by fluorescence measurements and preconcentration (solvent extraction or solid-phase extraction) has to be used prior to injection. This form of detection is also less specific than fluorescence and there is a greater possibility of chromatographic interference from other components in the sample. As with the case of phenols, the development of liquid chromatographic methods often stems from the difficulties encountered with analyses using gas chromatographic techniques. In many cases, this may be attributed to the polarities of the molecules (e.g. phenols and *N*-methylcarbamates), or their thermal labilities (e.g. *N*-methylcarbamates and phenylureas).

4.2.5 Immunoassay

The techniques described so far involve the use of complex laboratory equipment and often long pretreatment stages. Ideally, an analyst would like to achieve the required sensitivity and specificity with simpler equipment and without any pretreatment being required. Field analysis would also be desirable.

Part of the solution to this problem could be the use of *immunoassay* but as a separate test has to be designed for each analyte, it will never be the complete answer. Field kits (necessary apparatus, reagents and calibration standards for a specific number of analyses) are available in the μg/l range with an analysis time of 10–15 min. Laboratory kits are available for individual compounds in the ng l^{-1} range and are typically capable of handling 40 samples in a period of two hours. The methods commercially available include the analysis of individual pesticides (e.g. atrazine, carbofuran and paraquat), BTEX compounds (benzene–toluene–ethylbenzene–xylene(s)), total petroleum hydrocarbon (TPH), PCBs and PAHs, with the list continually expanding. Several of these methods are now approved by the US Environmental Protection Agency (EPA).

The use of these kits can be very simple, requiring little background knowledge. More thorough knowledge is necessary to understand the potential applications and limitations of the immunoassay process and the almost bewildering number of variations of the basic technique. Chemical and biological principles will both need to be understood and the techniques used sometimes seem to be more at home in a life sciences rather than a pure chemical laboratory. First of all, let us look at a simple method. Later, we will look at the background principles behind the method in an attempt to understand its particular merits.

4.2.5.1 Methodology

Most field and laboratory kits use a technique known as competitive ELISA (enzyme-linked immunosorbent assay). For laboratory analyses, reactions take place in the wells of a microtitreplate (Figure 4.11). These are plastic plates which contain typically 40, 48 or 96 wells for the simultaneous analysis of the samples and standards. An automatic scanner (microtitreplate reader) measures their light absorbance at specific wavelengths. This apparatus is commonplace in biomedical laboratories.

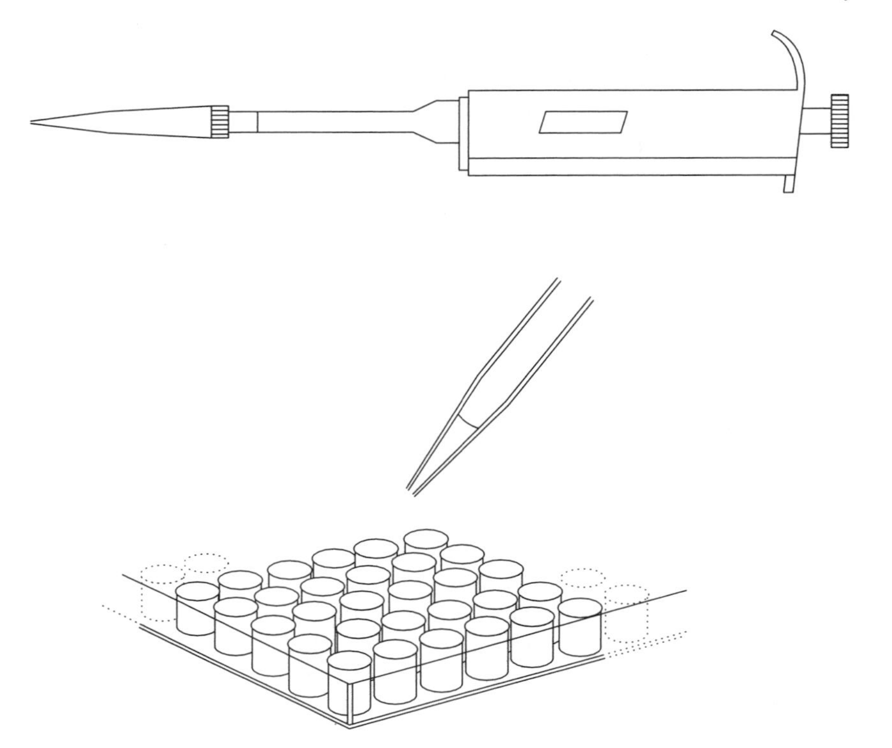

Figure 4.11 A micropipette and microtitreplate used in immunoassay.

The wells of the plate are filled with 100 μl sample or standards in duplicate. The reagents are then added. After a short period of time, the plate is then washed with water, further reagents are added and the plate is placed in an incubator at room temperature for a period of up to one hour. The absorption of light in each plate is then measured.

One design of field analysis kit includes individual pre-coated tubes, while another manufacturer has reagents attached to magnetic particles. The latter can be separated from the reagent and wash solutions by using a magnet, thus immobilizing the particles on the walls of the tube. Light absorbance is measured by a portable spectrometer.

The response curve is unlike any others you are likely to have come across, with a typical example being shown in Figure 4.12.

DQ 4.16

Comment on this response curve and suggest possible applications of this analytical technique.

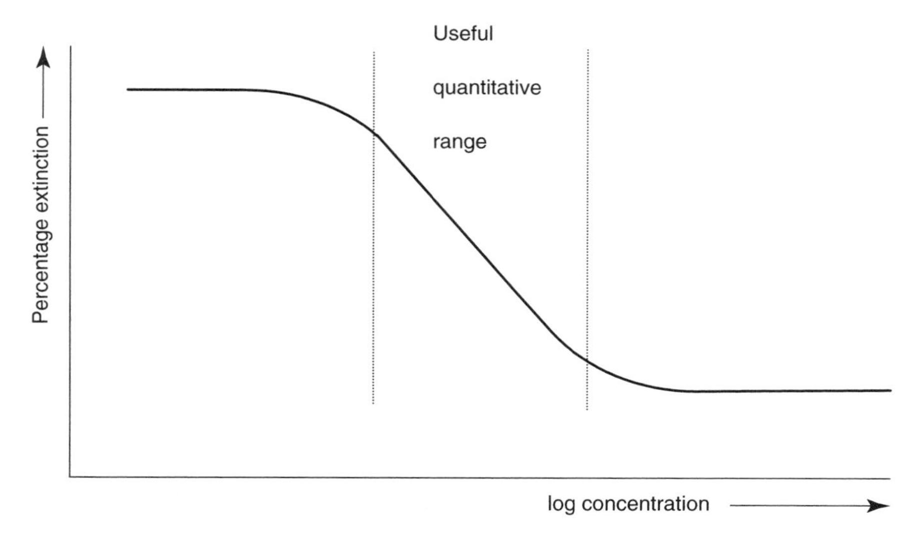

Figure 4.12 A typical response curve found in immunoassay.

Answer

You should notice that the range over which the response can be used quantitatively is quite limited, although it can be more easily used as an indication of whether the pollutant is present or absent. Application is often as a screen for potential pollutants. This will lower the number of samples requiring more expensive GC–MS or LC–MS analysis.

So, how does the technique work? The new terms which you will need to know are summarized in Table 4.3, while Figure 4.13 shows the steps involved in the immunoassay process.

The surface of the wells of the microtitreplate have been coated with an antibody. The samples and standards are introduced into the wells of the plate, each with a fixed amount of a labelled derivative of the pollutant. The derivative molecules compete with the pollutant molecules for binding to the antibody fixed on the plate. The amount of labelled derivative binding to the surface will be determined by the relative concentrations of the pollutant and derivative, and is inversely proportional to the concentration of pollutant originally in the sample.

For an ELISA analysis, the labelled derivative contains an enzyme moiety covalently bonded to the pollutant molecule. Other forms of immunoassay have different labels such as radioactive isotopes (*radioimmunoassay*) or fluorophores (*fluoroimmunoassay*). The washing stage removes the unbound pollutant and labelled derivative.

The enzyme is often used to catalyse a colour-change reaction. This results in the high sensitivity of the technique. One commercial kit for atrazine uses

Table 4.3 Common terms used in immunoassay

Antibody
A high-molecular-mass soluble protein produced within an organism which binds with the antigen by physical forces as part of the organism's natural defence mechanism. This will have a high specificity towards the antigen. The antibodies for the ELISA kits have been originally generated by innoculation of laboratory animals with the antigen.

Antibodies are often drawn as Y, which roughly indicates the shape of the molecule.

Antigen
Foreign material which can cause antibody production within an organism.

Clone
A group of genetically identical cells derived from a common parent by asexual reproduction.

Conjugate
A high-molecular-mass compound capable of producing an immune response. It is formed by covalently coupling a hapten with a soluble protein such as bovine serum albumin.

Hapten
A low-molecular-mass compound which will bind to an antibody even though the molecular mass is too low to induce the initial antibody formation. For environmental samples, this is generally the compound being determined in the immunoassay.

Immunoassay
An analytical technique involving the binding of antigens and antibodies.

Immunogen
A synthetic substance capable of inducing antibody formation in an immunized animal. For the immunoassays being discussed here, this is the conjugate.

Monoclonal antibody
An antibody formed by a single clone. This is effectively made up of identical molecular species and will be more specific to the antigen than the corresponding polyclonal antibody. These are produced by cell culturing techniques after selection from the polyclonal antibody.

Polyclonal antibody
An antibody formed in response to an antigen produced by several different clones. It is comprised of many molecular species, each with differing affinities and specificities to the antigen. Early assay kits used polyclonal antibodies, while recently developed kits may use monoclonal species.

---

the reaction of urea peroxide with tetramethylbenzidine, where a blue-to-yellow colour change occurs. After the reaction is stopped by a final addition of dilute acid, the absorbance is then monitored at 450 nm. The change in absorbance is inversely related to the concentration of the pollutant. The reaction is stopped after approximately one hour in most laboratory analyses. Field kits, however,

(a)

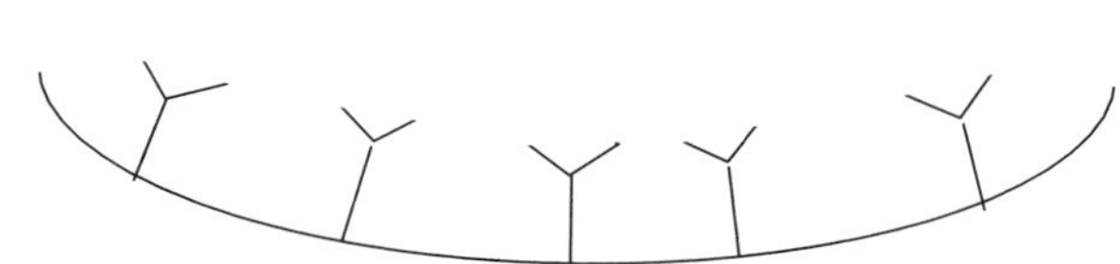

Microtitre plate pre-coated with antibody Y

(b)

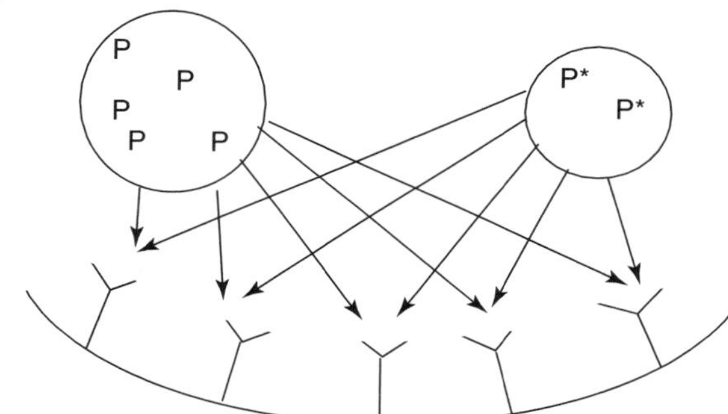

Pollutant P and labelled pollutant P* compete for antibody sites

(c)

Concentration of bound label P* is inversely proportional to the original concentration of P

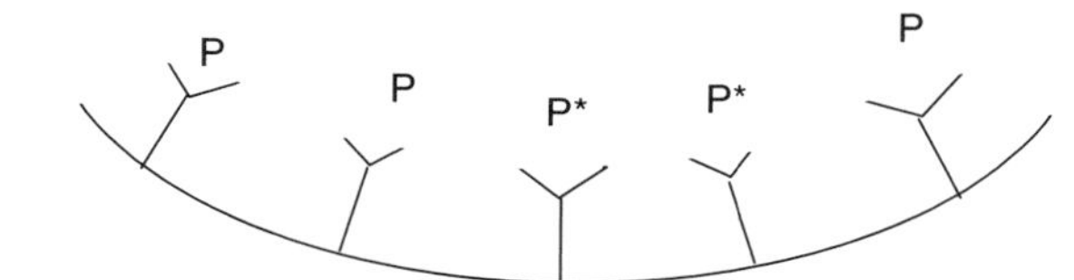

(d)

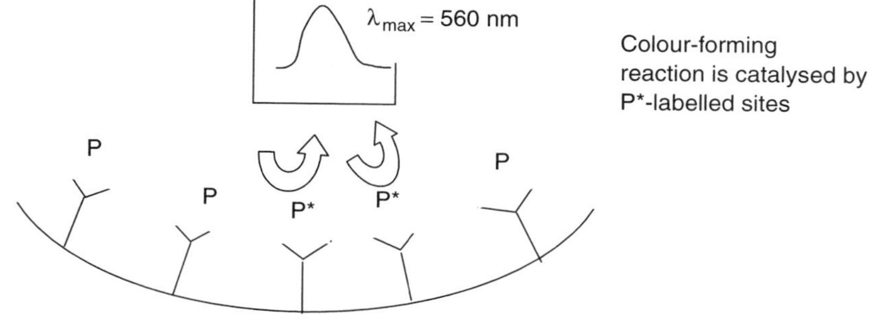

Colour-forming reaction is catalysed by P*-labelled sites

Figure 4.13 Steps in a typical ELISA.

stop the reaction after five to ten minutes to decrease the overall analytical time, but this will, of course, decrease the sensitivity of the technique.

4.2.5.2 Development of Tests and Implications for Analyses

An essential stage in the development of the kits is the production and isolation of the antibodies. The initial immune response can only be produced by molecules with an M_r greater than about 10 000, much larger than most pollutant molecules. Derivatives (conjugates) of the initial molecule (hapten) must first be produced by covalently bonding the latter to a carrier protein. Antibodies are generated by injection of the conjugate into a laboratory animal. After a few weeks, the antibody can be harvested from samples of the blood serum of the animal. Sufficient antibodies will be produced for several thousand kits. Monoclonal antibodies and genetically engineered antibodies are now becoming more common. These are single chemical reagents of a defined composition with constant specificity characteristics and can be mass produced.

As no two animals will produce identical antibodies, even 'identical' tests from different manufacturers will have to be considered as different analytical techniques and will need to be assessed separately.

The antibodies recognize molecules according to their molecular shape and bind at specific sites in the molecule. During the development process, tests have to be conducted to ensure the recognition sites ensure specificity for the compound being analysed. There is the possibility of 'cross-reactivity' with other compounds with a similar shape and functional groups. Cross-reactivity can, in fact, be used to advantage in some kits, e.g. the triazine pesticide test kit, which are designed to respond to groups of chemicals rather than to individual groups of compounds.

DQ 4.17

Identify features of an ELISA which means little pretreatment is necessary?

Answer

The use of antibodies makes the technique highly selective without pretreatment.

Enzyme catalysis gives the technique high sensitivity.

4.2.6 Spectrometric Methods

Often, a technique is required to measure the total concentration of a group of compounds, rather than individual concentrations. Such determinations include the analysis of the following:

- Total phenols
- Surfactants (total, anionic, cationic and non-ionic surfactants)
- Total hydrocarbon(s)

Visible spectrometry is often used for phenols and surfactant analysis after the formation of derivatives. Chromatographic methods have in the past not been used as they give *too much* information!

The simplicity of the method can be seen by the analysis of anionic surfactants using a 'Methylene-Blue' method. Under basic conditions, a salt is formed between the Methylene Blue and the surfactant and this salt can be extracted into chloroform. The absorbance of the extract is measured in the visible region (at 652 nm) and the concentration determined by comparison with a standard calibration curve.

Infrared spectrometry is used for the total hydrocarbon content. The hydrocarbons are extracted from the acidified water by using a non-hydrocarbon solvent (e.g. carbon tetrachloride) and the absorption measured at 2920 cm^{-1}, corresponding to the C–H stretching frequency.

SAQ 4.1

The Environmental Protection Agency (USA) lists 114 organic priority pollutants and suggests a purge and trap technique for volatile components and solvent extraction techniques for non-volatiles. Solvent extraction is used either under acid or base/neutral conditions.
Which technique could be used for the following compounds?

1. Toluene
2. Anthracene
3. 2,4,6-Trichlorophenol
4. Methylene chloride
5. Chloroform
6. 1,2-Dichlorobenzene
7. Phenol
8. Naphthalene
9. Hexachlorobenzene
10. Benzene

SAQ 4.2

The main use of the extraction methods discussed earlier in Section 4.2.2 is in the laboratory, often with the apparatus coupled as an integrated system to a chromatograph. At least two of the methods can be used for field sampling. You would take an extract back to the laboratory rather than an aqueous sample. Which techniques do you consider to be suitable in this case?

SAQ 4.3

Which GC column would be your initial choice for the analysis of the following:

(a) chlorinated pesticides in a natural water sample;
(b) volatile solvents in waste water;
(c) oil contamination in water?

SAQ 4.4

Which analytical methods could be used for the following compounds? Give your reasons for using the particular techniques.

- *N*-methylcarbamates
- Atrazine
- Phenols
- PAHs
- Malathion

4.3 Metal Ions

In this section, we will be predominantly looking at the analysis of metal ions found in the $\mu g\ l^{-1}$ to $mg\ l^{-1}$ concentration range. The only metals likely to be found above this range in natural waters are the four ions (i.e. sodium, potassium, calcium and magnesium) discussed earlier in Chapter 3. Of the remaining metals, iron, manganese and zinc can sometimes approach the $mg\ l^{-1}$ level, but other metal ions, if present, are likely to be at the lower end of this range.

Metal ions can occur naturally from leaching of ore deposits and also from anthropogenic (man-made) sources. Such sources include metal refining, industrial effluents, and solid waste disposal. Much solid waste, including power station fly ash, sewage sludge and harbour dredgings, contains significant concentrations of metal ions (up to 1000 $mg\ kg^{-1}$ total metal) which can leach into solution if in contact with water.

This area of analysis is currently dominated by techniques which can be grouped together under the general title of Atomic Spectrometry. The main individual techniques are as follows:

- Flame Atomic Absorption Spectrometry (Flame AAS)
- Graphite Furnace Atomic Absorption Spectrometry (GFAAS)
- Inductively Coupled Plasma-Optical Emission Spectrometry (ICP-OES)
- Inductively Coupled Plasma-Mass Spectrometry (ICP-MS)

These will be discussed, along with some other methods, in the following sections, showing the relative merits of each technique and their potential applications.

4.3.1 Storage of Samples for Metal Ion Analysis

You should by now be able to decide upon suitable sample containers and storage conditions applicable to most metals.

DQ 4.18

What are suitable sample containers and storage conditions?

Answer

(i) ***Polyethylene bottles are less likely to contaminate the sample with metal ions than glass bottles.*** *The only exception to the use of polyethylene bottles is for mercury analysis when glass bottles should be used. Mercury ions readily react with many organic materials.*

(ii) ***The sample should be acidified to minimize precipitation of metal ions.*** *A typical procedure is the addition of 2 ml of 5 mol l^{-1} hydrochloric acid per litre of sample.*

(iii) ***Scrupulous cleaning of bottles is important.*** *This usually includes an acid washing stage to ensure complete removal of trace metals. In the case of aluminium, the concern over contamination extends to glassware used in the subsequent analysis. You are often advised to pre-leach glassware with dilute nitric acid and to reserve glassware solely for aluminium determinations. Such a procedure would, in fact, be good practice for* all *metal analyses.*

4.3.2 Pretreatment

Most routine analyses require the total metal content of the sample, regardless of its chemical nature. Pretreatment can include evaporation to dryness and re-dissolution in acid, partial evaporation with acid, or digestion with acid at an elevated temperature for several hours. This is to dissolve suspended material and ensure that the metal is present as the free ion. The more modern techniques we will be discussing (GFAAS, ICP-OES and ICP-MS) are sufficiently sensitive and interference-free for the majority of samples to require no further pretreatment.

Most of the other analytical techniques require an extraction/concentration step for trace analyses. This may be a separate solvent extraction stage, as with flame AAS and some visible spectrometric methods, or may be a concentration stage in the analytical technique itself (Ion Chromatography and Anodic Stripping Voltammetry). Such a step can also serve to remove potentially interfering ions which may be present in far greater concentrations than the analyte.

The most common method proceeds with the formation of a neutral complex with an organic ion and extraction of this into an organic solvent (simple metal salts or ionic complexes would not extract). Up to a twenty times increase in concentration is possible in a single stage. The complexing agent used depends on the subsequent analytical procedure, and this will be discussed in the relevant sections. Other extraction/concentration methods include the use of chelating or ion-exchange columns. The metal ions are first held on the column, either by complex formation with the column packing material (chelating column) or by ion exchange. The ions are then eluted as a concentrated extract with an appropriate solvent, often an aqueous buffer.

4.3.3 Atomic Spectrometry

We will start with a discussion of the technique you are probably most familiar with – flame atomic absorption spectrometry (Flame AAS) – and then show how the other atomic spectrometric techniques overcome problems found in its use for trace metal analysis.

4.3.3.1 Flame Atomic Absorption Spectrometry

DQ 4.19

From your knowledge of this technique, draw and label a diagram of a flame atomic absorption spectrometer.

Answer

A schematic of a flame atomic absorption spectrometer is shown in Figure 4.14 below.

In this technique, a light beam of the correct wavelength to be specific to a particular metal is directed through a flame. The flame atomizes the sample, producing atoms in their ground (lowest) electronic energy state. These are capable of absorbing radiation from the lamp.

Although the equipment appears completely different from other forms of absorption spectrometry, the law by which the absorption of light is related to concentration is similar to that we have used already for the absorption of ultraviolet and visible radiation.

DQ 4.20

What is the law and what is its mathematical form?

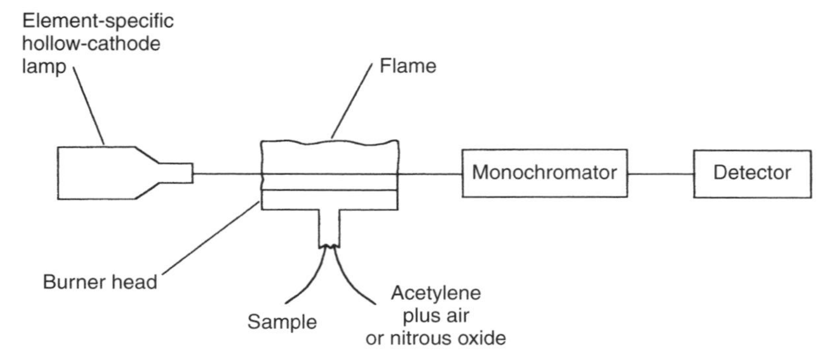

Figure 4.14 Schematic of a flame atomic absorption spectrometer (cf. DQ 4.19).

Answer

This is the Beer–Lambert law.

Check with the equations given above in Section 3.4.1 for the mathematical form.

The concentration range over which the law applies for flame atomic absorption spectrometry is usually 0–5 mg l^{-1}. Over the last three decades, atomic absorption spectrometry has dominated routine analysis of metal ions in aqueous samples at mg l^{-1} and higher concentrations.

DQ 4.21

From your previous knowledge of atomic absorption spectrometry, can you think of some of the advantages of this technique?

Answer

- *It is a rapid technique and can easily be automated.*
- *It is a simple method for routine use.*
- *Standard procedures are available for all metals.*
- *The analyses are generally free from interferences, while known interferences can easily be overcome.*
- *Apart from the pretreatment stages already mentioned, little or no sample preparation is needed for aqueous environmental samples.*

You may have included 'high sensitivity' within your list. I've left this out as I would like to discuss this further. Atomic absorption is indeed a sensitive technique and, if it is used for the more common ions discussed in Section 3.4 above, the water samples would have to be diluted before analysis. Magnesium is analysed by flame atomic absorption, often after sample dilution. If the technique is used for sodium or potassium analysis, lower-sensitivity absorption lines, rather than the highest-sensitivity lines, would be used in addition to diluting the sample. Atomic emission (flame photometry) is, however, the preferred technique for these ions.

In high-throughput laboratories, low-concentration samples (<1 mg l^{-1}) would normally be determined by the techniques described later in this section, particularly ICP-OES and ICP-MS. If these are not available, flame AAS can be used with sample preconcentration. This may simply involve partial evaporation of the acidified sample for zinc, iron and manganese analyses. Solvent extraction has been routinely used for other metals. Since atomic absorption analysis is relatively free from interference from other trace metal ions (i.e. the presence of other materials usually has little effect on the accuracy of the analysis), the extraction need not be highly specific to any one particular metal. In fact, it may be beneficial to be able to use a single complexing agent for several metals

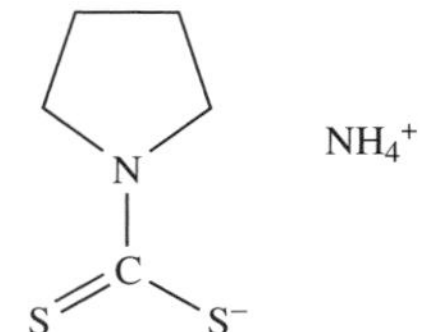

Figure 4.15 Structure of ammonium pyrrolidinedithiocarbamate.

since the extraction stage is the most time-consuming part of the analytical procedure. Ammonium pyrrolidinedithiocarbamate (APDC) (Figure 4.15) is often used as it forms stable complexes with most transition metals, if the pH is correctly adjusted. As an example, the optimum pH for lead extraction is 2.3. After extraction of the analyte in an organic phase, the organic phase is aspirated directly into the flame. The increase in the sensitivity is above that which is expected from the simple concentration factor. This is due to the increased aspiration rate resulting from the lower viscosity of the organic solvent in comparison to water.

You should be able to see a number of disadvantages of solvent extraction/flame atomic absorption, namely:

- It is very time-consuming.
- The sensitivity may still be insufficient for low-concentration metal ions.
- The risk of sample contamination is considerably increased.

To overcome these problems, other atomic spectrometric techniques have been applied to trace metal analysis.

4.3.3.2 Flameless Atomic Absorption

By replacement of the flame by other methods of atomizing the sample, the sensitivity can be increased sufficiently to remove the need for sample preconcentration. For most metals, this would mean the use of *graphite furnace atomization* (also known by the more general term '*electrothermal atomization*'), as shown in Figure 4.16, but, as we will see later, this is not the only method possible.

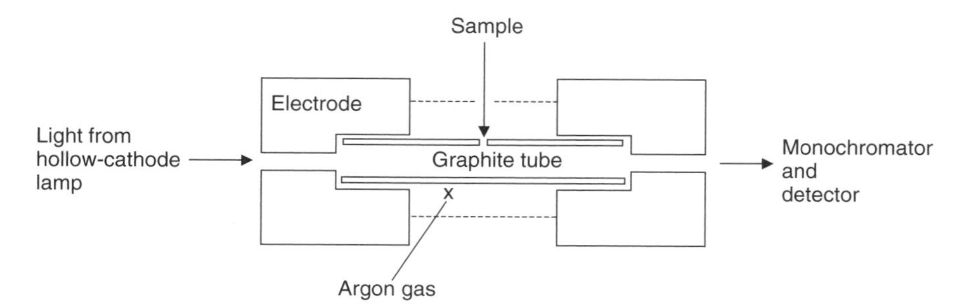

Figure 4.16 Schematic of a graphite furnace.

Graphite furnace AAS involves injecting a sample (up to 25 μl) into a small graphite tube (2–3 cm × 5–10 mm) which is heated in pre-programmed stages, as follows:

- Drying
- Decomposition
- Atomization

The absorbance of a light beam shone through the cell is measured during the atomization stage. The optimum temperatures and duration of each stage are metal-dependent, with a complete programme taking 2–3 min.

A comparison of flame and graphite furnace atomic absorption spectroscopies is presented in Table 4.4. As you can see from this table, the chief advantage of flameless AAS arises from removing the necessity of preconcentration of the sample. An extraction stage may still sometimes be necessary for complex samples in order to reduce potential interferences, as in the case of sea water analysis. One major source of error is background interference, which results from light scattering by solid particles within the beam. The scattering is highly dependent on wavelength, as follows:

$$\text{Scattering}\ \alpha \frac{1}{\lambda^4} \tag{4.5}$$

where λ is the wavelength of radiation.

The analytical wavelengths used for lead and cadmium are towards the far end of the available ultraviolet range and so analyses for these elements are highly susceptible to interference. Automatic background corrections should always be used for these elements. An analytical wavelength of 283.3 nm is also often preferred for lead, rather than the more sensitive 217 nm wavelength, as this lessens the effect of light scattering.

Background effects can also occur from other sources according to the analyte being considered, including the presence of thermally stable molecular ions. The

Table 4.4 A comparison of the advantages of flame and flameless (graphite furnace) atomic absorption spectroscopies

Advantages of solvent extraction/flame AAS	Advantages of graphite furnace AAS
• Simple technique	• Increased sensitivity ($\mu g\ l^{-1}$ concentrations)
• The solvent extraction stage can be used to remove potential interferences	• Decreased overall analytical time as the solvent-extraction stage is not usually necessary
• More readily available equipment	• Smaller samples required
• Shorter instrument time	• Unattended operation is possible
• Lower instrument cost	• Reduced risk of sample contamination

Table 4.5 Common background correction methods

Method	Feature
Deuterium lamp (continuum method)	A second absorbance measurement over a slightly larger wavelength range than the atomic absorption which gives the background reading
Zeeman	An intense magnetic field splits the absorption into magnetic components at slightly different wavelengths. Absorbance measurements with and without the magnetic field can be processed to correct for the background
Smith–Hieftje	A pulse increase in the lamp current removes the atomic absorption, leaving only the background

absorbance can be highly structured (e.g. narrow absorption bands within a much broader absorption). There are a number of methods available for background correction, as shown in Table 4.5. Due to the different principles on which the corrections are based, there may be advantages or disadvantages for each type according to the application. Revalidation may be necessary if a different technique is used to that specified in a standard method.

Other flameless atomization techniques can be used for specific elements. Inorganic mercury salts can be chemically reduced by using tin (II) chloride or sodium borohydride. The elemental mercury produced is then swept by a stream of nitrogen or air into a gas cuvette for absorption measurement in a modified spectrometer. Tin, lead and a number of metalloids (As, Se, etc.) can be reduced by sodium borohydride to volatile hydrides which are swept from the sample by a gas stream. Mild heating breaks down the hydrides to produce the elements in their ground states suitable for absorbance measurements.

4.3.3.3 Quantification

The major advantage of atomic absorption over other techniques is often stated as its lack of interference, particularly between metals. All that would appear necessary for quantification would be to use external standards to produce a calibration graph (see Section 3.4.1 above).

There are, however, a number of factors which will affect the accuracy of the analysis. One of these is chemical, typically where refractory salts are formed between the metal and an anion. These interferences are well known and usually described in instruction manuals accompanying the spectrometer. A typical concern for environmental samples is shown by the effect of phosphate on calcium, which decreases the absorption due to the formation of insoluble and refractory calcium phosphate. Similar problems can occur in the presence of sulfate and silicate ions. These problems can be overcome by adding a small quantity of release agent to each solution. A 10% lanthanum solution is often used. The lanthanum preferentially reacts with the phosphate. Alternatively, an

EDTA solution can be used. In this case, the EDTA complexes with the calcium, as shown in equation (3.23) above.

For more complex analytes, other factors may affect the accuracy. These include physical effects where the viscosity or surface tension of the solution is altered. Such properties will affect the aspiration of the solution into the flame and hence the measured absorbance.

The method of standard addition is often used to overcome this problem. A calibration curve is produced from a series of sample solutions which have been increased in concentration by adding known amounts of the metals ion being determined. This, of course, will increase the measured absorbance. The easiest way to achieve this for trace work where you do not wish to dilute the sample, is to add small volumes of higher-concentration standards so that the change in overall volume is negligible. The amount of metal added needs to be chosen so that the increase in absorbance is of the same order as that of the original sample. It is easy to see the principle. If you add, by chance, an amount of metal ion which will double its concentration, then the absorbance will double. It is perhaps a little less obvious to see how you calculate the unknown concentration from a series of additions. You do this by plotting a graph of the concentration increase against the absorbance. A linear plot should be produced, but, of course, the line does not pass through the origin. There will still be absorption from the metal ions in the untreated sample (i.e. the y-axis intercept). The concentration of the sample is found from the x-axis intercept, with the latter being the negative value of the sample concentration.

DQ 4.22

A series of solutions is made up by adding 0.1, 0.2, 0.3, 0.4 and 0.5 ml of a 10 mg l^{-1} lead standard to 100 ml aliquots of the unknown solution. The following results were obtained:

Volume of standard added (ml)	0	0.1	0.2	0.3	0.4	0.5
Absorbance	0.27	0.37	0.53	0.65	0.75	0.88

Plot a calibration graph from the above data and determine the concentration of the unknown.

Answer

Assuming that the volume remains constant at 100 ml, the concentration increases in the five solutions will be 10, 20, 30, 40 and 50 $\mu g\ l^{-1}$*, respectively. The graph produced is shown in Figure 4.17 below. The least-squares line is as follows:*

$$Absorbance = (0.012\,35 \times concentration) + 0.2694$$

which gives a concentration in the unknown of 21.8 $\mu g\ l^{-1}$ *lead.*

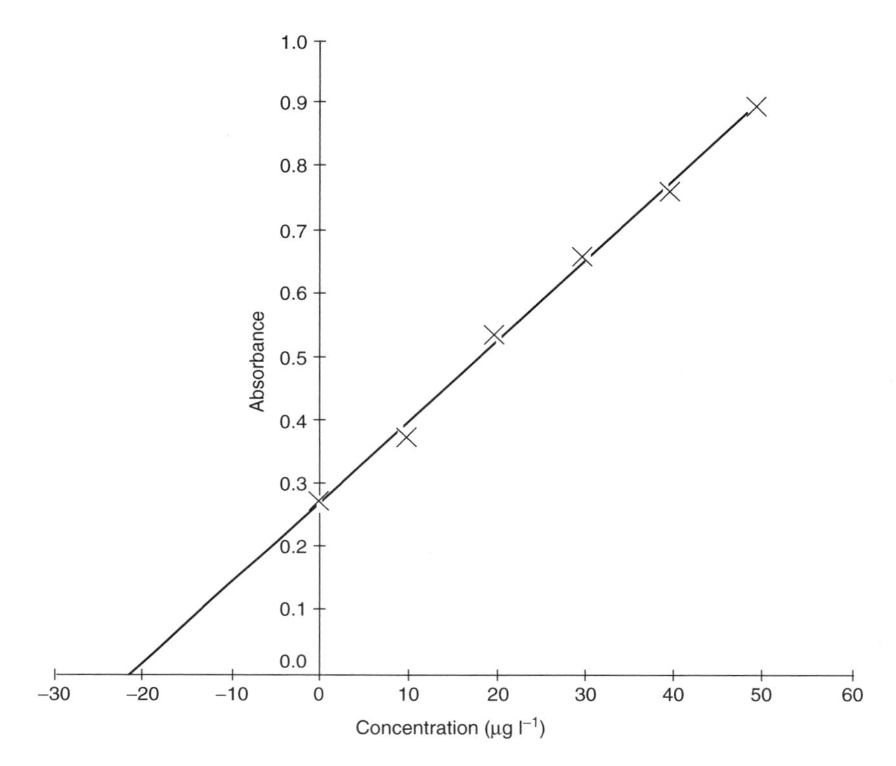

Figure 4.17 Typical calibration graph obtained in the 'quantification by standard addition' approach (cf. DQ 4.22).

Standard addition overcomes the problem of matrix effects as all the readings are taken from sample solutions with similar composition. The method does, however, produce particular problems of its own.

DQ 4.23

What problems can you see in this method?

Answer

(i) The concentration is determined by extrapolation. The latter will magnify any inaccuracies in the calibration line and is only possible if the calibration is linear.

(ii) Any error in zeroing the instrument for the analysis will be included as a systematic error in the analytical value produced.

(iii) You may be concerned about the volume of sample necessary to make up the standard solutions. Many modern graphite furnace spectrometers can, however, be programmed to perform the addition

automatically within the graphite tube by using microlitre quantities of sample and standard.

4.3.3.4 Inductively Coupled Plasma Techniques

Atomic absorption spectrometry has a number of disadvantages for use in analysing large numbers of samples of varying elemental composition and concentration.

DQ 4.24

What are the two major problems in the use of AAS for such samples?

Answer

AAS can only determine one element at any one time. The technique becomes slow and tedious for multi-element analysis. The variations in concentrations of the samples can be problematic as the linear range of AAS is very limited.

The development of the inductively coupled plasma (ICP) techniques for water analysis can be seen as an attempt to overcome these problems. At the same time, they maintain the advantages of graphite furnace AAS of being sufficiently sensitive not to require a preconcentration stage and also in not using flammable or explosive gases. This permits unattended, 24 hour, operation. In both methods, the sample is atomized in a plasma flame at 6000–10 000 K (Figure 4.18). This is generated by a flowing stream of argon which is ionized by an applied radiofrequency (RF) field.

Inductively Coupled Plasma-Optical Emission Spectrometry (ICP-OES). With this technique, the emission spectrum is monitored. Simultaneous ICP-OES can determine 60 or more elements at once by monitoring at pre-set wavelengths. This includes halogens and some other non-metals and metalloids, as well as metals. Sequential spectrometers, which are more common for water analysis, restrict themselves to a smaller number of elements, determined by the requirements of the analysis, measured in succession by rapid changes in the detection wavelength. The total analysis time is still fast, typically 5 s per element. A further advantage of ICP-OES is its wide dynamic range (approximately 10^5), which means that trace metals can be measured simultaneously with higher-concentration species.

In common with other emission techniques, there is the problem of spectral overlap from different elements, as an element will produce many more lines in its emission spectrum than in its corresponding absorption spectrum. The choice of the analytical wavelength is based on freedom from interference as well as sensitivity. For routine water analysis this problem has largely been overcome, with sensitive and interference-free lines being well documented. Quantitative

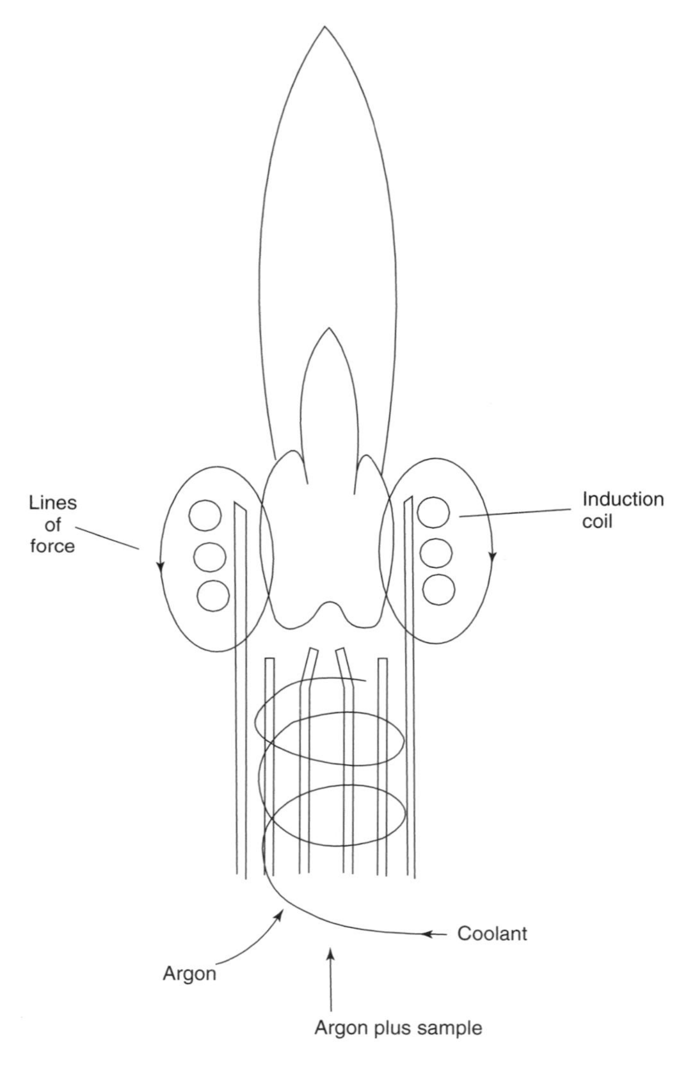

Figure 4.18 Schematic of a plasma flame unit used in ICP techniques for atomization of samples.

analysis can be performed by using external standards after first confirming that the chosen wavelength is free from interference. For quality control, monitoring can be at two wavelengths, which of course should produce identical results if there is no interference at either of the wavelengths.

For many years, the sensitivity of ICP-OES lay between those of the flame and furnace AAS techniques for most elements, thus making the technique useful for most, but not all, of the major components of water. Considerable effort was made to improve the sensitivity by changes in the spectrometer design. The major improvement lay in the relative position of the detector with respect to the plasma (Figure 4.19). Originally, this was at right angles to the plasma, so giving a short pathlength through the flame. The sensitivity is increased by 8–10× (i.e. to ca. furnace AAS sensitivity) by moving the detector to an axial position.

Problems which needed to be overcome included the effect of the plasma tail on the optics. One method of overcoming this is by diverting the plasma tail away from the optics by a radial flow of gas. Organic solutions cannot be used with axial flow detection (c.f. use of organic extracts with flame AAS). The linear range of the instrument is unchanged, but is moved to lower concentrations. As a consequence, both types of instrument are still in use today. Some instruments are capable of operating in either mode, with the axial configuration being reserved for applications needing higher sensitivities.

Inductively Coupled Plasma-Mass Spectrometry (ICP-MS). A more recent development is to use the inductively coupled plasma as an ion source for a mass spectrometer. For routine applications, this is usually a quadrupole spectrometer as is commonly found in a GC–MS system. The mass spectra of inorganic mixtures are simple in comparison to the more familiar organic compound spectra

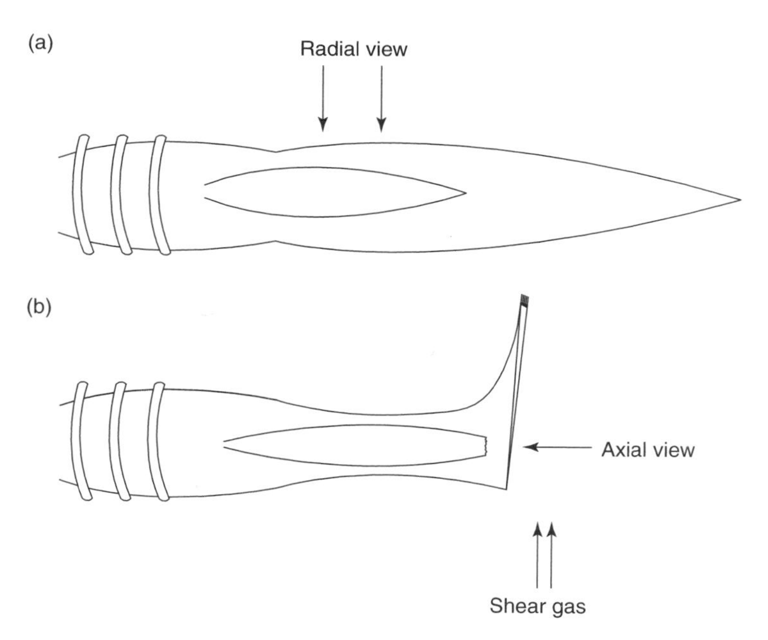

Figure 4.19 Relative positions of the detector in ICP-OES: (a) radial; (b) axial.

and fewer interferences occur in metal analysis. Although the technique is not strictly simultaneous – the ions being determined sequentially – determination of some 20 elements is possible within a period of 4 s. The sensitivity is slightly lower than that of graphite furnace AAS but is still sufficient to determine trace metal ions at below 1 $\mu g\ l^{-1}$ in aqueous samples. The linearity range is 6–8 orders of magnitude, according to the particular application.

Since each metal is determined according to its mass/charge ratio, there would seem to be, at first sight, little chance of interference in the technique. High concentrations of salts can cause deposition of solids in the instrument. Although there are few instances where there are problems with isotopes with the same mass/charge ratios, molecular ions can be formed, particularly with refractory elements. For example, $^{44}Ca^{16}O^+$ has the same mass/charge ratio as $^{60}Ni^+$, while $^{40}Ar^{35}Cl^+$ could interfere with $^{75}As^+$. Most of these problems can be eliminated by simple pretreatment of the sample to remove the potential interference before introduction into the ICP-MS system. Flow-injection techniques may be used to automate the process and may also be used to alleviate the problem of high salt concentrations if simple dilution is not possible. Avoidance of the use of hydrochloric acid for acidification will lessen problems with chlorine interference.

The use of mass spectrometric detection makes possible the use of the isotope dilution techniques discussed earlier in Section 4.2.3. The method would have to be applied to each isotope in turn and is time-consuming if large numbers of samples and elements have to be determined. As a consequence, for routine samples external standards are sometimes used after quality control checks to confirm the lack of any interferences. The standards may be matrix-matched to minimize problems caused, for instance, by viscosity differences. Standard addition (see Section 4.3.3 above) is also used.

Each development in atomic spectrometry has brought with it a significant increase in instrument capital cost. The cost of AA instrumentation is generally according to the following:

$$\text{Flame AAS} < \text{Furnace AAS} < \text{ICP-OES} < \text{ICP-MS}$$

In addition, ICP techniques have significantly higher running costs due to consumption of the argon necessary to generate the plasma. The advantages of ICP techniques are, however, so great that ICP-OES has for several years been the preferred technique for the high-concentration metal analysis in major water analysis laboratories, with ICP-MS being used for lower-concentration metals. Atomic absorption methods find a role in smaller laboratories where the sample throughput is insufficient to justify the additional capital and running costs of ICP techniques.

4.3.4 Visible Spectrometry

Until the widespread use of atomic spectrometric techniques, visible spectrometry was the most commonly used technique for metal ion analysis. Standard methods

Table 4.6 Some examples of colour-forming complexing agents

Metal	Reagent	Limit of detection ($\mu g\ l^{-1}$)
Iron (II)	2,4,6-Tripyridyl-1,3,4-triazine	60
Manganese	Formaldoxime	5
Aluminium	Pyrocatechol Violet	13

were developed for all commonly found metal ions. These methods use colour-forming complexing agents. Selectivity in the analysis is achieved in two different ways, as follows:

1. Solvent extraction is sometimes used. Chromium is analysed as the diphenylcarbazide complex after extraction into a trioctylamine/chloroform mixture. This gives a limit of detection of 5 $\mu g\ l^{-1}$ in the original sample. The complexing agent dithizone can be used for 17 metals. The selectivity is achieved by precise control of pH and the use of masking agents.
2. Alternatively, a colour-forming complexing agent can be used which is sufficiently sensitive and selective for use in the aqueous sample without extraction being necessary. Some examples are shown above in Table 4.6.

A number of these techniques have been adapted for use with portable colorimeters (e.g. iron, manganese, chromium and copper) and it is perhaps in this area that such techniques have the most widespread current usage.

It is useful to consider why such a well-established technique as visible spectrometry could become largely superseded by atomic methods:

1. Atomic methods are more rapid.
2. Although visible spectrometric pretreatment is generally simple when analysing relatively unpolluted water samples (rivers and lakes), they may become complex and time-consuming with more complicated samples such as sewage effluents.
3. Visible spectrometry is often affected by interference from other elements. This can be illustrated by the determination of iron using 2,4,6-tripyridyl-1,3,5-triazine. The concentration effects observed on a true value of 1.000 mg l^{-1} iron are shown in Table 4.7.

Nonetheless, visible spectrometry remains a frequently used technique and would be the method of choice when atomic methods are unavailable.

4.3.5 Anodic Stripping Voltammetry

A number of electrochemical methods are sufficiently sensitive to determine the low levels of metal ions typically found in the environmental water samples

Table 4.7 Concentration effects on the measured ion content (1.000 mg l^{-1}) of an iron sample

Additional ion	Effect
100 mg l^{-1} sulfate	-0.020 mg l^{-1}
2 mg l^{-1} cadmium	$+0.009$ mg l^{-1}
10 mg l^{-1} lead	-0.026 mg l^{-1}

without any separate preconcentration. Anodic stripping voltammetry (ASV) has found particular use in environmental analysis, where at least 19 metals can be analysed in this way.

The apparatus used consists of an electrolytic cell containing a working electrode (a mercury drop, or a thin film of mercury deposited on a glassy carbon electrode), a reference electrode and a counter electrode. A three-electrode system is used so that current and applied potential can be measured independently. This attempts to compensate for the change in potential drop due to the resistance of the test solution during the analysis. The latter would affect the measurement in a two-electrode system.

The sample is placed in the cell along with a supporting electrolyte (e.g. 0.1 mol l^{-1} acetate buffer at pH 4.5). Nitrogen or argon is bubbled through the solution to remove dissolved oxygen, which would otherwise interfere in the analysis. The working electrode is held at a small negative potential with respect to the reference while the solution is stirred. Reduction of the metal ions to the free metal occurs at the working electrode. Under controlled conditions of deposition time and stirring rate, the quantity of metal deposited on the electrode is proportional to its original concentration:

$$M^{2+} + 2e \longrightarrow M \tag{4.6}$$

After a predetermined time, the potential of the electrode is slowly changed in the positive direction. At specific potentials, depending on the metal and supporting electrolyte, each metal is oxidized and returned back into solution, as follows:

$$M \longrightarrow M^{2+} + 2e \tag{4.7}$$

This process is monitored by plotting the current change between the working and counter electrodes against the potential (Figure 4.20). The height of the peak in the curve is proportional to the concentration of the metal.

DQ 4.25

Can you see why this method does not require a separate concentration stage?

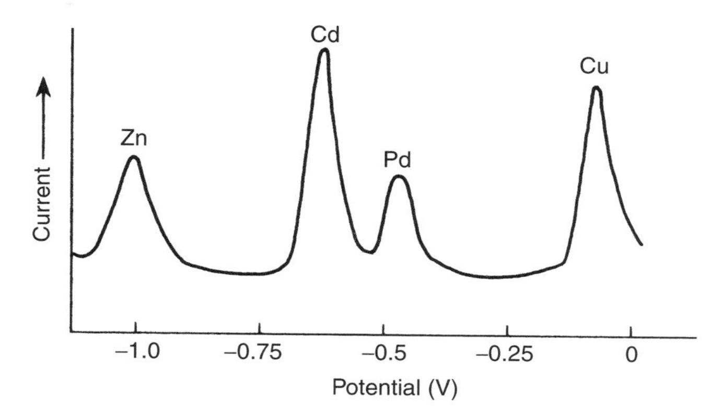

Figure 4.20 A typical anodic stripping voltammogram.

Answer

If you consider the experimental method, you will see that the first step, where the metal is being plated on the electrode, is itself a concentration stage – hence the high sensitivity of the technique with little pretreatment.

The supporting electrolytes necessary for individual metals are tabulated in standard texts. Electrolytes are often acidic, as their potentials become little affected by minor changes in the sample composition. Complexing agents (e.g. acetate) are sometimes included to stabilize particular oxidation states or to move the stripping potential of the metal away from potential interferences. Four metals of major environmental concern, i.e. copper, lead, cadmium and zinc, can, however, be analysed in a single scan by using the acetate buffer mentioned earlier. Quantification is by standard addition. As an electrochemical method, the technique determines only free metal ions in solution, plus, to some extent, loosely associated complexes.

However, anodic stripping voltammetry does have at least one disadvantage. The laboratory method is slow, with stripping times varying between 30 s and 30 min, and during such periods the apparatus is devoted to a single sample. Compare this with atomic spectrometry, where the instrumental time is only a few seconds per sample. The time taken for analysis has, however, been overcome with commercially available field instruments. These utilize disposable electrodes and microprocessor control which automatically takes the solution through the scanning cycle. Sample pretreatment is by addition of salts in tablet form. The conditioning breaks down any complexed forms of the metal and so the concentration output is of the *total* dissolved metal. A complete lead and copper analysis at the $\mu g\ l^{-1}$ level can be performed in ca. 3 min.

If the total metal content is required with laboratory apparatus, sample pretreatment is also necessary. This may range from simple acidification to UV irradiation in order to destroy any potential complexing agents. By performing the analysis with and without pretreatment, a measure of the free and complexed metal ions can be made, which would not be possible by using atomic spectrometry. This makes anodic stripping voltammetry a useful research tool, but because of its relative slowness in the laboratory, limited in its application for routine analysis.

4.3.6 Liquid Chromatography

As environmental waters invariably contain a large number of metal ions, and often at similar concentrations, you might think that liquid chromatography would be a frequently used analytical technique. Your argument might be that it should be possible to determine all of the metal ions present by using a single sample injection into the chromatograph. Atomic methods still, however, dominate metal analysis. Liquid chromatography only finds use in areas where atomic spectrometry is not ideal.

DQ 4.26

Re-read the earlier section on atomic absorption analysis of metals and decide areas in which the use of liquid chromatography may be preferable.

Answer

This isn't the easiest of questions to answer, although you may have found some of the following:

1. ***Complex Matrices***. *Extraction techniques are often necessary when complex samples are analysed by AAS in order to remove interfering components. This extends the time taken to perform an analysis considerably.*
2. ***Analysis of Mixtures of Uncommon Elements***. *AAS determines individual elements by using a different lamp for each one. Additional, and perhaps unsuspected, elements will not be detected. With a correct choice of column and eluent, these would be seen as additional peaks in a liquid chromatographic analysis. The need to change lamps for each element may also mean that AAS is a slower technique than chromatography for complex mixtures. You may also find that it is more difficult to obtain the hollow-cathode lamps necessary for AAS for the more uncommon elements.*
3. ***Quantification of Different Chemical Forms of the Ion***. *Later, we will be discussing the different chemical forms in which a metal can be found in the environment. In certain instances, ion chromatography can separate and quantify the chemical forms. Atomic absorption spectrometry is unable to distinguish the different species that may be present.*

If we extend our comparison to ICP techniques, then many of the perceived advantages may still hold. Interferences in complex samples may still be found. Sequential plasma emission spectrometers detect a limited number of elements and unsuspected elements may still be missed. Different chemical forms are not distinguished. ICP techniques are, however, more rapid for multi-element analyses.

Chromatographic methods using both dedicated ion chromatographs and conventional HPLC have been developed. The most sensitive method for transition metals in complex mixtures using a dedicated chromatograph is known as *Chelation Ion Chromatography*. This method involves the use of two preconcentration columns (Figure 4.21), and spectrometric detection after mixing with a derivatization agent, i.e. 4-(2-pyridylazo)resorcinol. The detection limits are 0.2–1 $\mu g\ l^{-1}$ with a 20 ml sample volume.

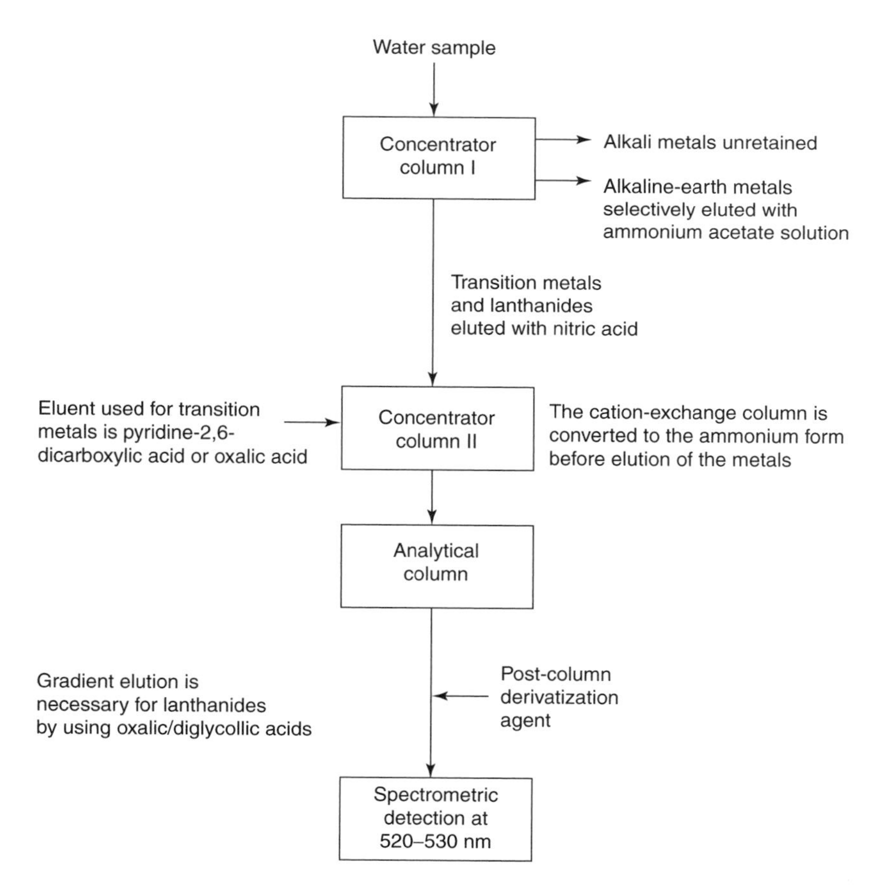

Figure 4.21 Schematic of a chelation ion chromatography system.

After acid digestion to ensure that the metal is present as the free ion, the sample is added on to the first column, eluted on to the second, and then on to the analytical column. Although we can say that ion chromatography removes the need for a *separate* preconcentration stage, you can see that preconcentration still occurs within the instrument as an identifiable analytical step. Separation of all common transition metals takes less than 14 min, with a typical chromatogram being shown in Figure 4.22. Furthermore, separation of the lanthanide elements requires less than 12 min to carry out.

A common application of chromatography for separate different chemical forms of the same element is for Cr^{3+} and $Cr_2O_7^{2-}$. Once again, the species are determined by visible spectrometry after derivatization. Pyridine dicarboxylic acid is reacted with the sample prior to the analytical separation to form a stable anionic complex with the Cr^{3+}(i.e. pre-column derivatization). Separation of the species employs an anion-exchange column. Diphenyl carbazide is added after elution from the column (post-column derivatization) to react with the $Cr_2O_7^{2-}$. Both species may be then detected at 520 nm.

A common method for trace metals when using conventional HPLC involves separation of the thiocarbamate complexes. Excess thiocarbamate (diethylthiocarbamate and pyrrolidine dithiocarbamate salts have both been used) is added to the sample and the transition metal complexes formed are concentrated by liquid–liquid or solid-phase extraction. The concentrated extract is injected into the HPLC system and separated by using a reversed-phase technique. Underivatized metal ions can be separated by reversed-phase ion-pair

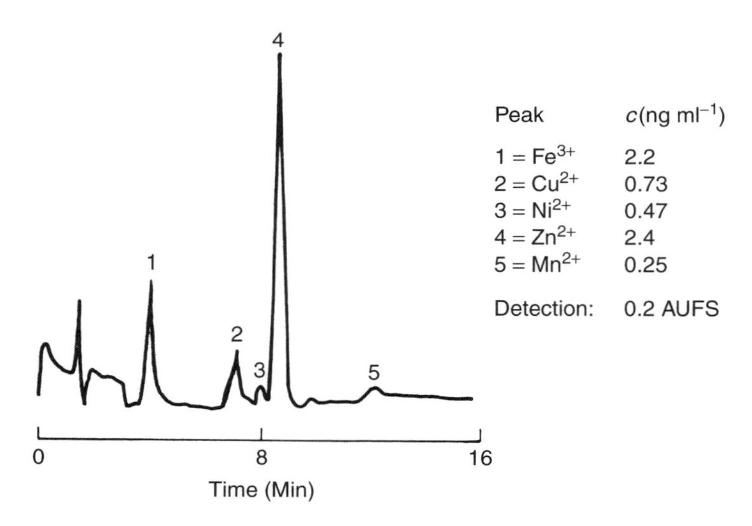

Figure 4.22 A typical chelation ion chromatogram, obtained for a sample of 32 g of sea water (Monterey Bay, CA). From Dionex Technical Note TN25, Dionex (UK) Ltd. Reproduced by permission of Dionex Corporation.

techniques or by using a cation-exchange column. Several detection methods have been used, including conductivity and post-column derivatization with 4-(2-pyridylazo)resorcinol.

4.3.7 Metal Speciation: A Comparison of Techniques

Speciation is defined as *the different physical and chemical forms of a substance which may exist in the environment.* When considering water samples, this includes not only the truly dissolved metal ions (as free metal ions or as complexes), but also colloidal forms of the metal and any metal contained within, or adsorbed on to, suspended particles.

DQ 4.27

What lead species do you think might exist in a river?

Answer

Some of the possible species are shown in Table 4.8 below.

Table 4.8 Possible forms of lead found in a typical river

Species	Example	Physical form
Free metal	Pb^{2+}	Solution
Ion-pair	$PbHCO_3{}^+$	Solution
Complexes with organic pollutants	Pb^{2+}/EDTA	Solution
Complexes with natural acids	Pb^{2+}/fulvic acid	Suspension
Ion absorbed on to colloids	$Pb^{2+}/Fe(OH)_3$	Colloidal
Metal within decomposing organic material	Pb in organic solids	Solid
Ionic solids	Pb^{2+} held within clays	Solid
	$PbCO_3$	Solid

Although I don't expect that you will have thought of all of these forms, I hope that you will now appreciate the great diversity of species which may be found. These include not only well defined ions and compounds, but also loosely bonded complexes and adsorbed species. The free metal ion often only comprises a small percentage of the total content. The interconversion between species is slow and for many purposes they can be considered as being distinct chemical forms.

For a number of metals, there may also be concern over the different organic derivatives in the environment. An example would be tributyl tin which has been used in anti-fouling paint formulations for ships hulls and its dibutyl and monobutyl degradation products. Other metals with important organic derivatives include lead and mercury.

The transport of each species in the environment will be different and they will also have different toxicological properties. As an example, let us consider the behaviour of metals within a stream and in the associated sediment. Any decaying vegetation will increase the metal loading in the stream water, since the organic acids produced as part of the decay process will form soluble coordination complexes with the metals. The toxicity of the stream water, however, may not be increased as much as you might expect. As a very general rule, metal complexes have lower toxicities than their corresponding free metal ions. If the metal has more than one stable oxidation state in water, there may also even be differences in behaviour between different oxidation states. For example, chromium in the form of $Cr_2O_7{}^{2-}$ has a greater toxicity than Cr^{3+}. It would appear that the $Cr_2O_7{}^{2-}$ ion can enter cells via routes which permit entry of the similarly sized $SO_4{}^{2-}$ ion. Such a route would not be possible for the positively charged Cr^{3+} ion.

Each of the analytical techniques described in this current chapter will 'respond' in a different manner to the species in solution – this is summarized in Table 4.9. If one of these techniques is included as part of a more lengthy analytical procedure, the pretreatment stages may also alter the species being analysed. Any filtration, for instance, will remove particulate matter.

Speciation may be investigated by taking advantage of the different responses of the analytical techniques, and the effect of pretreatment. The most common method is to perform several ASV analyses with different pretreatment stages. A simple two-step procedure would be to perform the analysis on samples with and without ultraviolet irradiation, thus giving a value for the free metal (or more precisely, the total ASV-labile content) and total metal content, respectively.

The complete chemical characterization of a sample would be exceedingly complex and time-consuming. When you remember that the total metal concentration may not be greater than a few $\mu g\ l^{-1}$, you will realize that you may also be reaching the detection limits of the available techniques. This aspect of

Table 4.9 Response of various analytical techniques to different metal species

Technique	Response
Atomic spectrometry	All the metal species in the sample, i.e. the total metal is determined
Visible absorption spectrometry	Free metal ions, plus ions released from complexes by the colour-forming reagent
Anodic stripping voltammetry	Free metal ions analysed, plus any ions released from complexes during analysis. The total is often referred to as the 'total ASV-labile content'
Liquid chromatography	Non-labile species can sometimes be determined separately
Gas chromatography	Organic derivatives can be determined separately

environmental trace metal analysis is currently of great interest and improved techniques are continually being reported in the literature.

DQ 4.28

A complete characterization of all species in a sample is a difficult and time-consuming procedure. Can you think of an alternative, and simpler, approach to species analysis to support investigations on metal transport and toxicology?

Answer

Rather than to attempt to determine each species individually, those with similar environmental transport or toxicological properties could be analysed as groups. A simple classification of metal species would be into 'organic-solvent-soluble' (neutral complexes and organometallic species) and 'organic-solvent-insoluble' (charged complexes and free ions) species. The first type would be transported in the environment and would accumulate in fatty tissues in a similar manner to neutral organic molecules (see Section 2.3 above), and the second type in a similar fashion to other ions (see Section 2.4 earlier) within the environment.

SAQ 4.5

Lengthy pretreatment techniques are often necessary with the analytical techniques described for both organic compounds and metals. Filtration, solvent extraction and chromatographic pretreatment are common methods. What could affect the precision of measurements for low concentrations of common pollutants ?

SAQ 4.6

An analytical technique for copper, lead, cadmium and zinc in water, as described in the chemical literature, involved dividing the filtered sample into five aliquots. The treatment of the aliquots (prior to ASV analysis) was as follows:

(i) Strong chemical oxidation and UV irradiation.
(ii) No pretreatment.
(iii) Weak chemical oxidation.
(iv) Passage through a chelating resin then UV irradiation.
(v) Extraction using an organic solvent, and UV irradiation of the aqueous phase.

Which species are determined in each aliquot?

Summary

Components present at trace ($\mu g\ l^{-1}$) levels can have a major affect on water quality if they can bioaccumulate in organisms or have a high degree of toxicity. These components usually fall into the two categories of organic pollutants and metal ions. Instrumental methods for the determination of the components have been discussed, along with the necessary extraction and pretreatment steps. The predominant instrumental technique for organic components is gas chromatography, whereas atomic spectrometric techniques are the most frequently used methods for metal ion analysis.

Chapter 5

Analysis of Land, Solids and Waste

Learning Objectives

- To understand the problems of sampling and pretreatment of solid samples prior to the analytical determination of organic compounds and metal ions.
- To apply your knowledge of these problems to the analysis of:
  - plants and animals
  - soil
  - contaminated land
  - waste and landfill disposal sites
  - sediment
  - sewage sludge
- To appreciate newer extraction techniques for solids and to be able to assess their role in relation to the longer-established techniques.

5.1 Introduction

This chapter introduces you to methods for the sampling and extraction of solids. These are necessary stages prior to completion of analysis by the instrumental techniques which have already been discussed for water samples. They may, in fact, be the most difficult part of the analysis. Modifications which are needed for the final analytical stage are described where these are appropriate. Be sure that you can remember the principles of the transportation of pollutants discussed in Chapter 2 as these are necessary to understand the relevance of the analysis.

DQ 5.1

Which solids do you consider to be of importance for the study of the environment? What specific analyses would be relevant?

Answer

1. *Animal and plant specimens*
2. *Soils and contaminated land*
3. *Waste and landfill waste disposal sites*
4. *Sediments and sewage sludge*
5. *Atmospheric particulates*

We shall consider all of these in detail in the following discussion.

1. *Animal and plant specimens*
These are directly of interest since the toxic effect of a compound is proportional to its concentration within the organism. The investigations would also be relevant to species further along the food chain when determining the environmental pathway of the pollutant (see Section 2.3 earlier).

Plants and animals may also be used as indicator organisms to monitor pollutants found in lower concentrations in the wider environment. As an example, heavy metal pollution in sea water is often monitored by analysis of seaweed rather than by direct analysis of the water. Remember, however, that you have to balance the advantage of the higher concentration in the living organism with the disadvantage of the more complex analytical matrix.

The effect of pollution on living organisms can sometimes also be investigated by monitoring levels of naturally occurring constituents of the organism. The effects of acid rain on trees, for instance, include a decrease in the concentration of the alkaline-earth ions in the leaves.

2. *Soils and contaminated land*
Soils are complex materials comprising the following:

- Weathered rock
- Humus
- Water
- Air

Soils provide nutrients for plants, as well as providing physical anchorage and support for growth. Nutrients include nitrogen (in the form of nitrate and ammonia), phosphorus (in the form of orthophosphate) and trace metals such as copper, iron, manganese and zinc. However, not all of the ionic material within soil can be extracted by plants. Some is too strongly bound within the soil structure. Although a total analysis of soil is sometimes performed, the more frequent need is the determination of the **available** ionic material. The transport of material in the soil is influenced by the acidity or alkalinity of water in the soil structure, and so soil pH is frequently monitored. A further common analysis is for metal ion or organic contamination of the soil, resulting, for instance, from the misuse of pesticides, dumping of waste material or deposition of pollutants from the atmosphere.

The UK definition of contaminated land is 'land which represents an actual or potential hazard to health or the environment as a result of current or previous use'. The contamination may come from a time when there was little environmental legislation. Problems from more modern sites can occur when there have been spillages. There is current emphasis on reuse of land from former industrial 'brown-field' sites for new applications rather than using unspoilt 'green-field' sites; hence the necessity for assessing the potential problems.

3. *Waste and landfill waste disposal sites*

One of the major ways of disposing of waste is through *landfill*. The waste may contain a large amount of biodegradable material with low concentrations of toxic components (e.g. municipal waste) or it may contain high concentrations of more hazardous material (hazardous or special waste). The United Kingdom disposes of 27 million tonnes of municipal waste by landfill each year.

When a disposal site is full, the site is then capped, usually by a layer of clay. Although physically stable, the site is not chemically stable and its composition can continue to change over many years. Emissions are in the form of liquid leachate which can affect groundwater and also gases and vapours. The sites will need monitoring for these emissions during construction and for many years afterwards as the composition slowly stabilizes.

4. *Sediments and sewage sludge*

We have already discussed how both organic compounds with low water solubility, and metal ions, tend to accumulate by adsorption on to fresh water or marine sediments (see Sections 2.3 and 2.4 above). Analysis of the higher concentrations found in sediments may be an easier task than analysis of the surrounding water. Such analysis may be of the adsorbed species only, or it may be of the total sediment. The latter is often fractionated according to particle size prior to the analysis. This is important as the adsorption of pollutants can often be related to the available surface area, which in turn is related to the particle size. The latter also affects the mobility of the sediment and the possibility of ingestion by marine organisms.

Sewage sludge is the inert material produced as the end-product of the sewage treatment process. The material is sometimes spread on land as a soil conditioner or may be disposed of by incineration, or dumped as a waste product. The greatest concern over this material is the metal content, which may be as high as 1000 mg kg^{-1} total metal in some sludges.

5. *Atmospheric particulates*

An important route for the transport of inorganic salts and neutral organic compounds is via atmospheric particulate deposition (see Section 2.2 above). Typical determinations include the elemental analysis of particulate material and the analysis of the organics adsorbed on to the particulate surface. Once again, particle size may have to be determined.

5.2 Common Problem Areas in the Analysis of Solids

In the following sections, we will discuss the analysis of most of the solids mentioned above, but will reserve the discussion of particulates until we have looked at other components of the atmosphere. Despite the diverse nature of these solids, their analyses have common problem areas. After making a number of general comments on sampling, pretreatment, extraction and analytical determination common to all of the solids, we will then look at the analysis of each solid type in detail.

5.2.1 Sampling

The concentrations of the analytes may vary widely from sample to sample, both on a local scale (e.g. adjacent soil samples taken from a field) and on a larger scale (adjacent fields). 'Identical' plants can possess different levels of contamination. For example, leaves on the windward side of trees are exposed to atmospheric pollution to a greater extent than those on the leeward side. The variation, or inhomogeneity, has to be reflected in the number of samples to be taken. The subsequent analysis could be of each individual sample, or after combining a large number of samples and sub-sampling to obtain an average concentration. Sampling positions have to be chosen with care. For monitoring exercises involving large areas, a careful and planned choice of sampling sites should be made. Some of the common methods used for this are presented in Table 5.1.

Invariably, other considerations (e.g. geographic features or the availability of suitable plant specimens) prevent a regular sampling pattern. Areas where there is the possibility of specific contamination from other sources should be avoided. It would be easy, for example, to sample soil for general pollution by monitoring close to roads, because of ease of access and proximity to the ideal grid location, while at the same time forgetting the localized pollution from the road traffic. Whenever possible, duplicate samples should be taken at each location to assess local inhomogeneity. If monitoring is related to a point source into the atmosphere (e.g. factory emissions), then consideration should be given to the prevailing wind direction, with an increase in the number of sampling positions in the area of highest likely contamination. In addition, prevailing water currents should be taken into account for marine discharges.

Table 5.1 Some common methods used for determining sampling locations

Method	Description
Regular grid	The area is subdivided by using grids, and samples are taken at fixed locations on the pattern. Often a square grid is used, although sometimes the pattern may be more complex, e.g. the 'herringbone pattern' discussed later in Section 5.5.1
Stratified random sampling	The site is divided into small areas of equal size and a given number of sampling points are chosen at random within each area
Unequal sampling	This is used if a preliminary investigation shows areas of high concentration. The size of the sampling areas and number of samples is decided so as to define the contaminated areas more accurately

Control samples should also be taken at points remote from the area under investigation and an effort should be made to match the control site as closely as possible to the sample site. If, for instance, a factory discharge located in an urban environment is being monitored, then the control site should be an urban site where there is no possibility of similar contamination.

The sampling of contaminated areas may involve drilling or digging. There is potential of contamination of the sample from the equipment used and also from the difficulty of its cleaning between sampling. The components of the sampler which are in direct contact with the sample would need to be inert and would be typically constructed of stainless steel or high-density polyethylene. Care is also necessary to prevent more indirect contamination from, say, pumps or motors being used for the sampling or from elsewhere on the site.

5.2.2 Pretreatment

This can include the following:

- Washing of sample
- Drying
- Grinding/homogenization

These procedures are often deceptively simple. It is easy to forget that most samples are biologically or chemically active and even washing, prolonged warming, or storage at room temperature may change their composition. In addition, some analytes may be thermally unstable, volatile, or even photolytically unstable. Contamination or analyte loss is also possible at each stage of the analytical procedure.

5.2.3 Extraction of the Analyte

This could involve any of the following processes, depending on the analysis being undertaken:

- Extraction into an organic solvent – semi-volatile organics
- Vapour-phase extraction – volatile organics
- Ashing and subsequent dissolution – elemental composition
- Extraction using aqueous solutions – 'available' ions

Many of the analytes are very common contaminants (some, such as DDT, are classed as 'universal' contaminants) and may be present in the extraction agents or adsorbed on to the apparatus. High-purity reagents, specific to the analysis, should be used. For instance, 'pesticide-free' grade solvents are available from some manufacturers. A blank sample should be included in the analytical scheme to monitor contamination during the analytical process, but the preparation of a 'blank' sample may in itself be difficult for universal or near-universal contaminants.

Methods which have been in use for many years are first described in the following sections, followed by some newer techniques. The latter are instrumental methods which attempt to improve on the classical methods by being more easily automated, with decreased extraction/digestion times and often decreased solvent consumption. All allow a higher throughput of samples. The methods and instrumentation described are now commercially available and are rapidly becoming accepted within the standard methods.

5.2.4 Sample Clean-up

The extraction process from solid samples will almost inevitably lead to the co-extraction of other compounds. This would include not only low-molecular-mass compounds (perhaps other pollutants), but also high-molecular-mass materials. The term 'lipid' is often used for such naturally occurring high-molecular-mass organic species which can be extracted into organic solvents. Clean-up is vital before chromatographic analysis is carried out. As well as the techniques discussed earlier in Chapter 4 (e.g. column chromatography and solid-phase extraction), additional techniques can be used for lipid removal. *Gel permeation chromatography* separates compounds according is their molecular size. The equipment used is either a classical low-pressure column or a preparative-scale high-pressure liquid chromatograph. The method can be used for a broad range of pollutants, regardless of any knowledge of their detailed chemical structures. You should contrast this with adsorption columns (see Sections 4.2.2 and 4.2.3 above), which need a prior knowledge of the polarity of the compounds to determine suitable separation conditions. Chemical methods which destroy the lipid by saponification (a typical reagent is 20% potassium hydroxide in ethanol), or by dehydration or oxidation (concentrated sulfuric acid), are often too harsh for many pollutants (e.g. organophosphorus pesticides), but can be used for some organochlorine pollutants.

5.2.5 Analytical Determination

The instrumental methods of analysis generally follow the procedures outlined in the previous two chapters.

DQ 5.2

What did we find were the most common procedures for the analysis of organic compounds and for metals in aqueous samples?

What extra considerations would you think necessary for the analysis of solid extracts?

Answer

Most organic materials would be analysed by using gas chromatography. Atomic spectrometric methods (AA, ICP-OES and ICP-MS) are the most common techniques for metals.

Due consideration would have to be taken of the possible interferences from other components that may be extracted. For gas chromatographic analysis, this could take the form of careful assessment of the clean-up techniques prior to the chromatographic analysis (see Section 4.2.2 above). High-resolution columns may be a necessity. If atomic absorption techniques are used for metal analysis (you may find that flame atomic absorption often has sufficient sensitivity), then background corrections will be required (see Section 4.3.3 above).

5.2.6 Quality Assurance and Quality Control

In Section 4.2.3 above, we discussed the addition of standards prior to the pretreatment of water samples to determine the recovery efficiencies. A similar procedure could be used to determine extraction efficiencies from solid samples.

DQ 5.3

What is the problem with direct addition of a standard to a solid sample to determine the extraction efficiency?

Answer

There is no guarantee that the extraction efficiency of the standard will be the same as the analyte. The later may be so strongly bound within the solid structure that it would be incompletely extracted. The internal standard may be less loosely bound, particularly if the extraction takes place immediately after addition, and so would be more easily extractable.

There is no easy way around this problem, although allowing the standard to equilibrate with the sample for several hours would be a good practice to adopt. There is always the possibility that extraction may not be complete for the analyte even if the standard does indicate complete extraction. During the validation of new analytical techniques for solids, there is often a confirmatory analysis of certified reference materials. Such reference materials are chosen to correspond as closely as possible to the samples being determined.

Many of the applications of solid analysis involve sampling in areas of heavily contaminated land. In these circumstances, constant quality checks are needed to confirm that no sample contamination has taken place. This is carried out by including in the analysis scheme blank samples introduced at each stage of the sampling procedure (see SAQ 2.5 above). Any positive result from a blank would indicate contamination at that stage.

SAQ 5.1

A monitoring exercise is planned for lead deposited on soil close to a busy roadway. What sampling positions would you select?

5.3 Specific Considerations for the Analysis of Biological Samples

We will discuss first the sampling and extraction of components from plant material, and later consider the differences in approach which may be necessary in the case of animal tissues.

5.3.1 Sampling and Storage of Plant Material

The sample may be foliage, roots, or the whole plant. A single species should be sampled, with each specimen, if possible, being at a similar stage of maturity. If foliage is being sampled, the minimum sampling height should be such that there is no possibility of contamination by upward splashing from the soil, assuming that the species is tall enough! The maximum height is often determined by practical considerations. A suitable sample size is often 500–1000 g. The sample may be stored under refrigeration for a few days if it cannot be analysed immediately.

5.3.2 Pretreatment

5.3.2.1 Washing

Even a simple procedure such as washing may extract the analyte. Patting the sample dry by using a paper tissue can result in contamination of the sample with trace metals. It is consequently often preferable to avoid washing altogether if suitable clean samples can be found. Soft brushing may be an alternative. The cleanliness of samples is particularly important for trace metal analysis where the concentration may be higher in the surrounding soil than in the plant specimen.

Some pollutants may have been deposited from the atmosphere on to the leaf surface. If you are studying uptake of the pollutant by the plant, then this would have to be removed by washing. If, however, you were studying transfer of the pollutant along the food chain, then this should be included or determined separately. Dioxins, for instance, are not taken up by plants, but can enter the food chain by deposition on leaves which are then eaten by herbivores.

5.3.2.2 Drying and Homogenization

DQ 5.4

Which two factors do you consider will determine the temperature and drying time of a biological sample?

Answer

The temperature and duration of drying must be a balance between too low a temperature over a protracted period promoting biological activity and too high a temperature over a shorter time period leading to loss of volatile components.

A typical drying procedure would be to blow a current of dry air over the sample for a period of up to 12 h. The temperature should not be in excess of 50°C. Alternatively, the sample may be freeze-dried, i.e. deep-freezing the sample, reducing the pressure and removing the water by sublimation.

DQ 5.5

Due to the risk of potential losses caused by drying, why do you think such treatment is necessary at all?

Answer

I suspect your answer will be so that the analytical result can be referred to a dry weight. There is, however, no reason why you could not calculate the dry weight on a second sample. Drying the samples lessens the possibility of change due to biological activity. A second advantage is that homogenization of the bulk sample, necessary if sub-samples are to be taken, is made easier if the sample is dry.

Homogenization of the dried sample is often by the use of a high-speed grinding mill. Care should be taken, once again, to ensure that you are not introducing contaminants during the grinding process.

5.3.3 Extraction Techniques for Organic Contaminants

If present, organic contaminants are likely to be in the $\mu g\ kg^{-1}$ concentration range or below. The simplest method for the extraction of organics is to shake a sub-sample with an extracting solvent (e.g. hexane or petroleum ether for neutral organics) and to leave the two phases in contact for several hours. An alternative method is to use Soxhlet extraction. The apparatus employed for this is shown in Figure 5.1. In this method, fresh solvent is continuously refluxed through the finely divided sample contained in a porous thimble and a syphon system removes the extract back into the refluxing solvent. The net effect is continuous extraction by fresh solvent. A typical extraction takes about 12 h and uses about 300 ml of solvent. The technique is only applicable to analytes which can withstand the reflux temperature of the solvent.

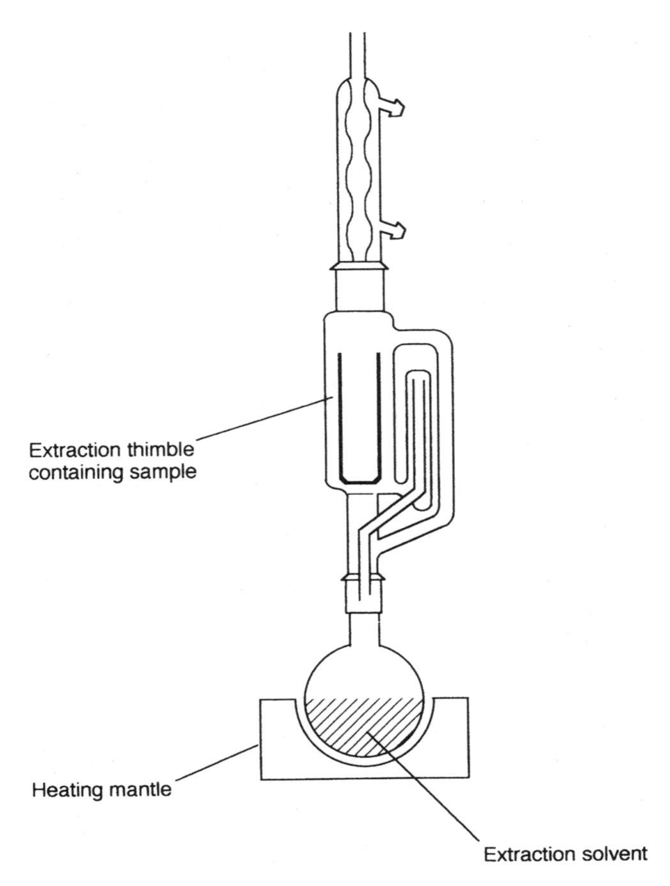

Figure 5.1 Schematic of a Soxhlet extraction system.

DQ 5.6

The use of hexane or petroleum ether as extraction solvents assumes a dried sample. What problems could you foresee if for any reason the sample was not dried?

Answer

The two solvents suggested are immiscible with water and would easily form an emulsion. In some cases, it may be difficult for the solvent to penetrate the sample.

A desiccant is often mixed with the sample during the extraction – sodium sulfate is commonly used for this purpose. The solvent can be modified by the addition of a polar solvent such as acetone. Alternatively, the extraction solvent can be changed completely to a solvent which is miscible with water. Acetonitrile is often used in this case. You must, of course, be certain that the solvent is still appropriate for the extraction of the analyte. If the sample will not allow the solvent to penetrate the structure, conditioning the sample with a polar solvent which is miscible with both water and the extraction solvent may overcome the problem. Isopropanol has been used for this purpose.

5.3.4 Ashing and Dissolution Techniques for Trace Metals

Trace metals are likely to be in the mg kg^{-1} concentration range. The concentrations, however, will vary from species to species and throughout the growing season. In order to extract metals, the organic matter is decomposed by dry or wet ashing.

Dry ashing consists of heating the sample in a muffle furnace, typically at 400–600°C for 12–15 h. The resulting ash is then dissolved in dilute acid to give a solution of the metal ions. Inaccuracies can arise both from volatilization of metals and the retention of metals in a insoluble form in the crucible.

Wet ashing consists of heating the sample with oxidizing agents to break down the organic matter. A typical procedure would be heating first with concentrated nitric acid, followed by perchloric acid. Alternative combinations include sulfuric acid/hydrogen peroxide and nitric/sulfuric acids.

An advantage of wet digestion is lower losses from volatilization (due to lower temperatures and liquid conditions), but it can give rise to higher metal blanks from impurities in the acids. Great care has to be taken with methods involving perchloric acid. This acid, in the presence of metals, has a tendency to detonate on drying! Small sample sizes should always be used and the liquid in the flask should never be allowed to dry out.

5.3.5 Analysis of Animal Tissues

Although the above discussion is specifically for plant analysis, much is also applicable to the analysis of animal tissue if certain differences in the sample types are first recognized. Animal samples are more liable to decomposition than plant samples and should be preserved by freezing below 0°C. Typical concentrations in specific tissues for both metal ions and neutral organic compounds can be in the mg kg^{-1} range.

Organic compounds are extracted without drying the sample. Often, the bulk sample is homogenized in a blending mill with water and sub-samples are then taken from the slurry for extraction. Remember also the previously described method of extraction from moist samples by inclusion of a solid drying agent. An alkaline digestion stage is also often included before organic extraction to break down any fatty tissue. Metals are once again extracted after wet or dry ashing.

SAQ 5.2

A method sometimes suggested to ensure complete extraction of the organic pollutants from biological samples is to repeat the extraction a number of times with different solvents. What disadvantage would this technique have for the subsequent stages of the analysis?

5.4 Specific Considerations for the Analysis of Soils

5.4.1 Sampling and Storage

Soil composition may vary greatly over a small area. We have already discussed how samples will have to be taken from a number of locations to obtain a suitable average composition. There will also be differences in composition according to the depth of sampling.

For polluted soil, you should take into account the source of the pollution and its mobility within the soil, which in turn may depend on the soil composition and pH. Some pollutants deposited from the atmosphere are generally quite immobile and will remain within the surface layer. Lead contamination from vehicle exhausts decreases rapidly (within a few cm) with depth. Dioxins similarly remain in the top layer of soil, with the molecules becoming strongly bound within the soil structure. Other pollutants may be more mobile. Samples should be cooled or frozen for transportation to the laboratory.

Some typical samplers are shown in Figure 5.2. If the soil is disturbed (e.g. by ploughing), samples should be taken from the whole of the disturbed area. If the investigation is concerned with possible uptake by plants or crops, then the sampling should be over the whole depth that the root system penetrates (which

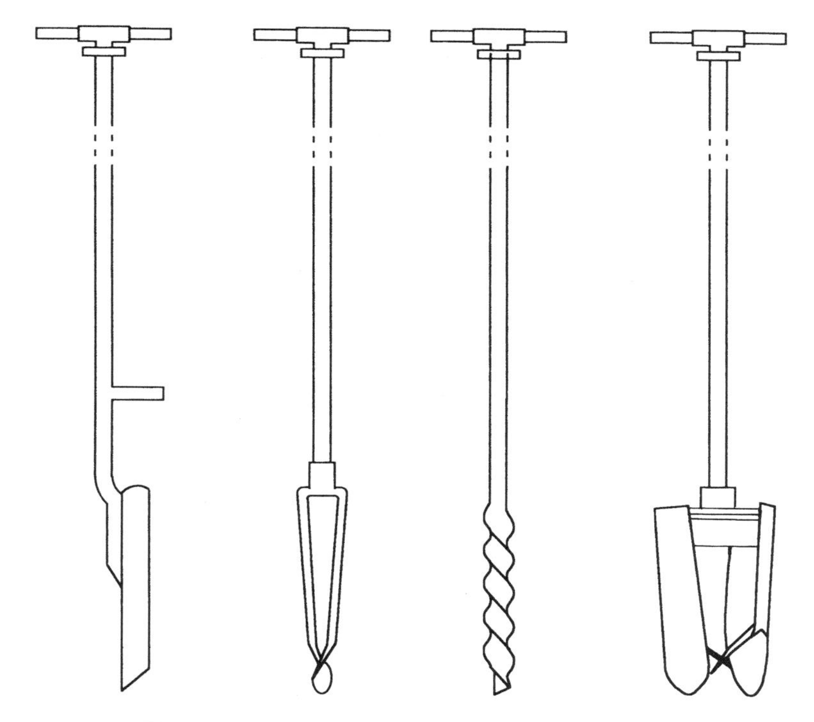

Figure 5.2 Examples of some typical soil samplers.

may be greater than the depth of any ploughing). For landfill sites, samples should be taken over the complete depth of the landfill.

DQ 5.7

Suggest reasons why there will be changes in the composition of soils throughout the year.

Answer

You may have included the following:

- ***Biological activity**, which will consume nutrients, will be greatest in spring and summer.*
- ***Rainfall**, which may leach out components, will vary throughout the year.*
- ***Pesticides and fertilizers** will only be applied at specific times in the year.*

5.4.2 Pretreatment

5.4.2.1 Drying

Until you remember the great amount of microbial activity in a typical soil sample, it may not seen obvious why as much care has to be taken with drying soil samples as with biological samples. Soil is often dried by equilibrating with the atmosphere at room temperature (under certain circumstances this may be raised to 30°C) for not less than 24 h. Under harsher conditions, the levels of available nutrients may change (this is particularly the case for phosphorus, potassium, sulfur and manganese), while nitrogen-containing compounds interconvert. The last problem is so great that analyses for nitrogen compounds should always use 'field-moist' samples.

5.4.2.2 Grinding

The drying process leaves the soil in large aggregates which need to be broken down into the constituent particles whose sizes range from 2000 μm for the coarse sand component, to less than 2 μm for clay. This is carried out by using a mortar and pestle, after sieving to 2 mm to remove pebbles and other large particles.

5.4.2.3 Sub-sampling

This is always a problem with solids, as any agitation tends to fractionate mixtures according to particle size. The smaller particles tend to fall below the larger particles. Standard methods are well established to overcome this problem, with the simplest being *the cone and quartering technique*. In this, the total sample is formed into a symmetric cone. The latter is then divided vertically into segments and alternate quarters are combined, with the remaining half being rejected. The process can be repeated successively until the required sub-sample size is produced.

5.4.3 Extraction of Organic Contaminants

Organic contamination is typically in the $\mu g\ kg^{-1}$ concentration range. Extraction of low-volatility compounds closely follows the techniques used for plant samples, which were described earlier in Section 5.3.3. There may, however, be the possibility of contamination by volatile or semi-volatile organic compounds. For low concentrations, the soil sample is mixed with water and the analysis then follows the procedures used for volatile organics in Section 3.2 above (e.g. purge and trap or head-space techniques). For higher concentrations, the organic is first extracted from the soil with methanol and this methanol extract is then added to water. Purge and trap or head-space analysis then follows.

5.4.4 Extraction of Available Ions

The concentrations of available trace metals and available nutrients would be expected to be in the mg kg^{-1} range. We will divide our discussion into a general method for the extraction of ions and then look at additional procedures which are needed for the special case of the measurement of nitrogen availability.

First of all, we need to know what is meant by the term 'available'. The complex structure of the soil acts as an ion exchanger for both cations and anions, where the simple ions are held to the soil by ionic forces. These ions may only be released into water from the soil by being displaced by ions of a different metal. This release will be dependent on the soil type and the chemical composition of the extraction water.

Analytical procedures for available ions attempt to reproduce the environmental conditions by a suitable choice of extracting solution. The procedure is simply to shake the soil with the extracting solution for a fixed period, typically one hour. A range of extracts have been used for this, including ammonium acetate solution, dilute acetic acid, dilute hydrochloric acid and EDTA solution, in order to mimic local conditions. This has led to problems in comparison of results with other laboratories where a different extractant may have been used and consequently a different proportion of the ions released. Once in solution, the ions can be analysed by the methods discussed above in Chapters 3 and 4.

5.4.4.1 Nitrogen Availability

The nitrogen species found in soil are as follows:

- Organic nitrogen
- Nitrate
- Nitrite
- Ammonia (free ammonia and ammonium ion)

Only the last three constitute the readily available nitrogen. Organic material is subject to microbial decay which will release nutrients over a period of time and so a measurement of organic nitrogen should also be included in the scheme.

The ionic forms of nitrogen can be extracted with potassium chloride solution. A subsequent reduction with, for example, titanium (III) sulfate, quantitatively converts all the ions into ammonia which can then be determined by standard methods.

DQ 5.8

What is the standard method for ammonia analysis?

Answer

This is by increasing the pH of the solution with sodium hydroxide, distilling the ammonia into boric acid, and titrating with standard acid (see Section 3.4.4 earlier).

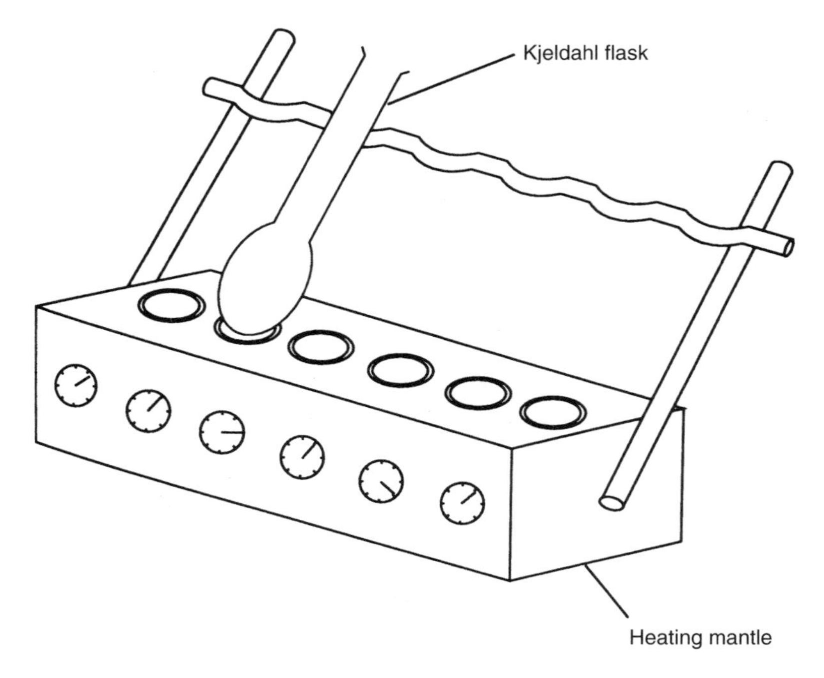

Figure 5.3 Schematic of a Kjeldahl apparatus.

Organic nitrogen is measured after a preliminary conversion to ammonia. This is achieved by boiling with concentrated sulfuric acid for several hours (Kjeldahl method). Potassium sulfate is added to raise the boiling point of the sulfuric acid, along with a catalyst. Selenium or mercury are often used for this purpose. A schematic of the typical apparatus employed is shown in Figure 5.3.

5.4.5 Dissolution Techniques for the Determination of Total Metal Concentrations in Soil

The *available* metal concentration as determined in the last section is only part of the *total* metal concentration in the soil. The total concentration analysis is occasionally required for environmental investigations. Extreme conditions have to be used to dissolve the soil, such as dissolution in hydrogen fluoride/perchloric acid mixtures, or fusion with an alkaline flux (e.g. sodium carbonate) and subsequent dissolution in dilute acid. Once in solution, the metal concentration can be determined by the standard techniques described previously in Chapters 3 and 4.

5.4.6 Determination of pH

Although soil contains water as an essential constituent, it is, of course, predominantly a solid. Since pH can only be defined as the hydrogen ion concentration *in solution*, then the pH of a soil sample is the pH of water in equilibrium with that particular soil.

DQ 5.9

This definition of soil pH gives a hint to a potential difficulty in this seemingly simple analysis. What is this problem?

Answer

The water is in equilibrium with the soil. Any change of the conditions (even adding more water) can alter the equilibrium, and hence the pH.

At the very least, a thick paste of soil and water is necessary to measure the pH. The added water should be such that there is minimum disturbance to the solution equilibrium. A salt solution (potassium or calcium chloride) is often used to form the paste, and this is then left for one hour for the equilibrium to be re-established.

SAQ 5.3

Which instrumental methods would be used for the analysis of the following species in soil extracts:

- Potassium
- Calcium
- Magnesium and trace metals
- Available phosphorus (present as orthophosphate)?

5.5 Specific Considerations for the Analysis of Contaminated Land

Since 1990 in the UK, there has been a legislative requirement for local authorities to compile registers of contaminated land. A similar concern is reflected through legislation in many other countries. Contamination may have penetrated deep into the soil and can be in any form – inorganic or volatile/semi-volatile organic. The organic material may be stable for long periods or may rapidly biodegrade. There could be a problem of groundwater contamination, and the volatility of the contaminants may also produce atmospheric problems. A potential hazard of building houses on contaminated land containing biodegradable material is the accumulation of methane in the house from the anaerobic decomposition of the material. In certain cases, this could lead to an explosion! A second problem is illustrated by the recent local discovery (in Tyne and Wear, UK) that a number houses had been built on the site of old dry-cleaning works. In this case, it is possible that the soil had been contaminated with chlorinated solvents from the previous industrial activities.

DQ 5.10

Look at the UK definition of contaminated land given above in Section 5.1. What effect does this definition have on likely analytical schemes and subsequent data interpretation?

Answer

The definition includes the phrase 'actual and potential hazard'. Compounds or ions may be present in such low concentrations that they do not present a hazard unless they are known to bioconcentrate (see Section 2.3.1 earlier). Some estimation of the total quantity of the contaminant on the site is also necessary.

The physical and chemical forms of the materials (i.e. speciation) will need to be determined as they will affect whether a material will be released under given environmental conditions. Consider the difference in the toxicities of chromium (III) and chromium (VI) discussed earlier in Section 4.3.7.

The definition is based on 'hazard to health'. Consideration has to be taken of the potential migration of the compounds and the location of target organisms or vulnerable sections of the environment. Sampling and analyses should then be concentrated on this route (see Section 2.6 above).

The interpretation of the data with respect to whether there is a 'actual or potential hazard to health' depends to a large extent on the end use of the land. The same analytical data can be interpreted in different ways according to its future use!

5.5.1 Steps in the Investigation of Contaminated Land

The first stage in sampling contaminated land is background paper research into the history of the site to decide any potential problems. A simple walk-over site inspection could then be made for visible signs of pollution. This can then be used to decide what analysis may be necessary and to determine a sampling strategy.

DQ 5.11

Why is the determination of sampling sites a particular problem for contaminated land?

Answer

It may not be obvious how to define the area in which contamination has occurred. A small number of sampling sites may, in fact, miss areas of pollution. Sampling can also often be complicated by the variety of solids which may make up the 'land' on an old industrial site. This could be soil, sand, shale, brick, remnants of concrete buildings and other industrial waste.

Particular care should be taken where there is likely to be small areas of relatively high concentrations of contamination ('hot spots') that the sampling scheme will be the most suitable. It is quite possible that contamination from a single source on to sloping land is in the form of a ribbon from the source. If a regular grid sampling strategy is used (see Table 5.1 above), a herringbone-type grid is sometimes suggested as this would be less likely to miss ribbon contamination than the common square-grid pattern (Figure 5.4).

At this stage, simple surface tests could be performed, sampling with trowels or with one of the soil samplers shown earlier in Figure 5.2. You should remember the potentially corrosive nature of many industrial contaminants and all tools should be either PTFE-coated or made of stainless steel.

A number of field monitors have been developed for rapid site assessment to lessen the need for expensive laboratory analysis. These include the following:

(i) *Immunoassay test kits*. Did you notice when we were looking at water immunoassay kits (see Section 4.2.5 above) that many were for industrial contaminants? Much of the development for these kits has been for contaminated land analysis. The kits include simple apparatus to extract the contaminant from the soil into solution. Immunoassay of the extract then follows.

(ii) *Portable X-ray fluorescence (XRF) spectrometers*. XRF is a method for elemental analysis which has the great advantage that it can directly analyse

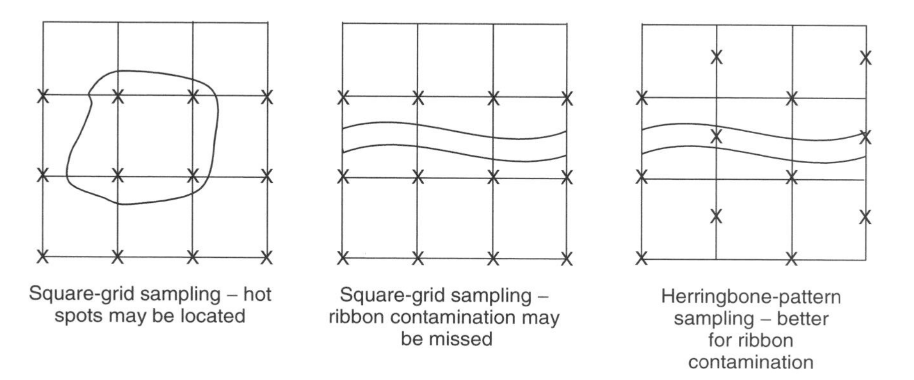

Figure 5.4 Some examples of sampling strategies employed for localized contamination.

solids as well as liquids. Within chemical analytical laboratories, it has been used much less than atomic techniques (see Section 4.3.3 above) due to lower accuracy, often attributable to strong matrix effects. Simplified instruments can be made portable and give readings in the mg kg^{-1} concentration range. Lead analysis is a typical application. The theory and instrumentation of XRF will be described later in Section 7.4.1.

(iii) *Monitors for specific groups of compounds*. These are based on spectroscopic properties which can identify the groups without separation. UV fluorescence monitors can be used for PAHs (c.f. Section 4.2.4) and infrared absorption for hydrocarbons (cf. Section 4.2.6). Some other uses of IR for environmental analysis will be discussed later in Chapter 6.

(iv) *Gas monitors*. These include portable gas chromatographs, methane gas analysers and direct-reading instruments for the analysis of individual gases. These will all be described below in Chapter 6.

5.5.2 Sampling, Sample Storage and Pretreatment

After the preliminary survey, laboratory analysis may still be necessary. Sub-surface samples could be taken either by using trial pits (or trenches) or by drilling. Each method has its own particular merits.

Pits are typically up to 3–4 m deep and are often dug by mechanical excavation. Solid samples are taken from the solid material in the scoop of the digger. The exposed soil layer structure can give additional information for site assessment. This approach is not ideal for water sampling as there is a large potential for contamination or mixing with solids. Gas or vapour samples would not normally be taken.

Drilling may cause fewer problems, although pollutants may still transfer between soil layers during the drilling process. A number of solid sampling methods are possible, including using an open-tube sampler which is attached to the end of the drill rods. This could be driven into the ground by using either an hydraulic jack or with a drop-weight system.

One of the problems of the sampling is that it is unlikely that there would be anyone present with expertise in trace chemical analysis and it is an exceedingly difficult task to avoid sample contamination when using large-scale equipment. There will be limited possibilities for cleaning equipment on site. Available water for cleaning may itself not be pure. Ideally, the process could include detergent and steam cleaning. Detailed cleaning protocols are necessary to minimize possible contamination.

Water samples should be extracted from the boreholes after the land and groundwater has re-equilibrated from its construction, perhaps after 1–2 weeks. Sampling uses either pumps or bailers. A bailer can be a simple container with the sample entering via a ball valve (Figure 5.5) or it can be a more complex design with valves closed by a messenger dropped down the attachment line.

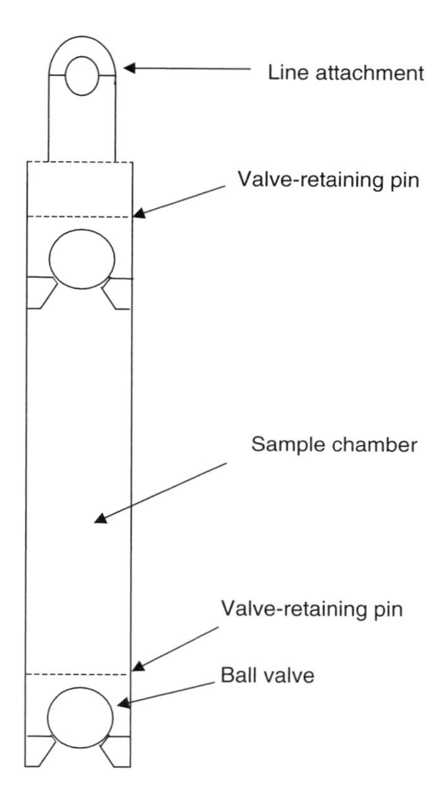

Figure 5.5 Schematic of simple bailer.

This allows sampling to take place at specific depths. Bailers are relatively cheap and so a different one can be used for each new borehole. This would remove the possibility of cross-contamination. However, only small sample volumes (a few hundred millilitres) can be extracted at any one time.

If pumps are used for water sampling, they need to be of low enough flow to minimize mixing of the water. Such mixing would increase the suspended solid content. Suction pumps are only useful to a depth of 20–25 ft and may strip out dissolved gases or volatile organics.

A conventional borehole takes 6–8 h to drill. A number of techniques (direct-push probes, such as 'KVA' samplers, and also 'geo-probes') are now available which can reduce the time for drilling samples by a factor of at least 10. This leads to a reduction in the overall cost of the analysis. Their more rapid use can mean that a larger number of samples may be taken. In comparison with established techniques, there can be a lower reproducibility of the results and their main use is in the rapid screening of sites.

Solid samples are stored either cooled or frozen and should be kept in the dark. Water samples should be preserved according to the principles already discussed (see Sections 3.2, 4.2 and 4.3 earlier). If gas sampling is by adsorption on to a solid (see Section 6.2 below), the loaded adsorption tubes should be stored cooled.

Pretreatment is similar to that used for uncontaminated soils (see Section 5.4 above) with the possible inclusion of a preliminary separation of the samples into solid types. Porous materials such as brick are efficient absorbents of contaminants. Extraction and analysis would then follow the procedures described in Chapters 3 and 4, and elsewhere in this present chapter. Gas analysis will be described later in Chapter 6.

SAQ 5.4

Place in order, starting with the most sensitive, the sensitivity of the following sites to past contamination. What effects from any contamination would you consider important to monitor? Note that some of the effects may be applicable to more than one of the sites.

(a) Public gardens and parkland
(b) Residential area with gardens and allotments
(c) Car park with a hard surface
(d) Commercial site

5.6 Specific Considerations for the Analyses Involved in Waste and its Disposal by Landfill

5.6.1 Types of Waste and their Disposal

Each country has its own definitions but, in general, there are at least two major categories. These represent non-hazardous waste and waste which requires additional care in its disposal. Non-hazardous waste could include municipal waste, commercial waste and some industrial wastes. A typical composition of municipal waste (defined in the EU to include waste from households and other waste of similar nature or composition) in a developed country is shown in Table 5.2. Hazardous waste, according to the United Nations Environmental Programme (UNEP) definition, is 'waste which is likely to cause danger to health or the environment, either alone or in contact with other waste'. Compounds could be, for instance, chemically active, toxic, explosive or corrosive. Special waste in the UK is defined as 'waste which is or may be so dangerous or difficult to dispose of that special provision is required for its disposal'. This waste can come from a large variety of sources, not just from industry. Have a look in your home to find cleaning agents and garden pesticides which may come under this category.

Table 5.2 Typical composition of municipal solid waste in a developed country

Component	Content(%)
Paper	25–40
Metals/glass	7–25
Food waste	6–18
Yard wastes	5–20
Plastics	4–10
Textiles	0–4

Although these types of waste are often referred to as 'solid waste', individual components of the waste may also be semi-solid sludges or liquids. This presents a major analytical problem, not only for monitoring of the disposal site itself, but also of materials prior to disposal.

Much of the waste is disposed of by landfill. There are differences in the landfill process, not only according to the type of waste but also from country to country. The disposal method will, of course, affect the nature and rate of emissions into the environment. A *containment landfill site* is constructed with a natural (clay/shale) or synthetic lining. Any release of liquid (*leachate*) into the wider environment would be expected to be low but would still need to be monitored. A *co-disposal site* is where several types of waste are mixed in order to promote natural processes of degradation. It is assumed that by the time any water leaves the site it will be environmentally acceptable. Most modern sites would include some form of containment to minimize the release and have collection systems to treat the leachate. *Entombment* is where the waste is stored in a relatively dry form so there is slow degradation of the waste with time, with the site taking up to 50 years to stabilize. Monitoring of the disposal site is necessary, not only as the site is being filled, but must be continued as the site matures. Regulations may require this monitoring for many decades. Analyses could be on site, using mobile laboratories, or using samples taken to remote laboratories.

DQ 5.12

Which analyses would you consider important for the environmental monitoring of landfill sites?

Answer

Environmental concern would be centred on monitoring compounds escaping from the site. This could be either as aqueous leachates or as gaseous emissions. Analyses may be for major components and their inter-reaction products, as well as for trace compounds. Analyses is also necessary for any materials (e.g. solids, sludges or liquids) being added to the site. You may consider the major environmental concern over waste is for hazardous or special waste but you should realize that ***problems can also arise from non-hazardous waste****. As organic waste breaks down, a leachate can be produced with an extremely high BOD. There could also be hazardous trace components.*

The high biodegradable component of municipal waste is of current concern within the EU and strategies are currently being formulated for its minimization. Waste disposal may be the most common form of entry into the environment for some trace pollutants. Polychlorinated biphenyls are in use for electrical insulation and have largely been in sealed units. It is only when they are incorrectly discarded that there is any major escape into the environment.

5.6.2 Sampling and Storage

Much of the routine monitoring is for liquid and gaseous emissions and sampling locations may be a permanent feature of the site. These are shown for a typical containment site in Figure 5.6.

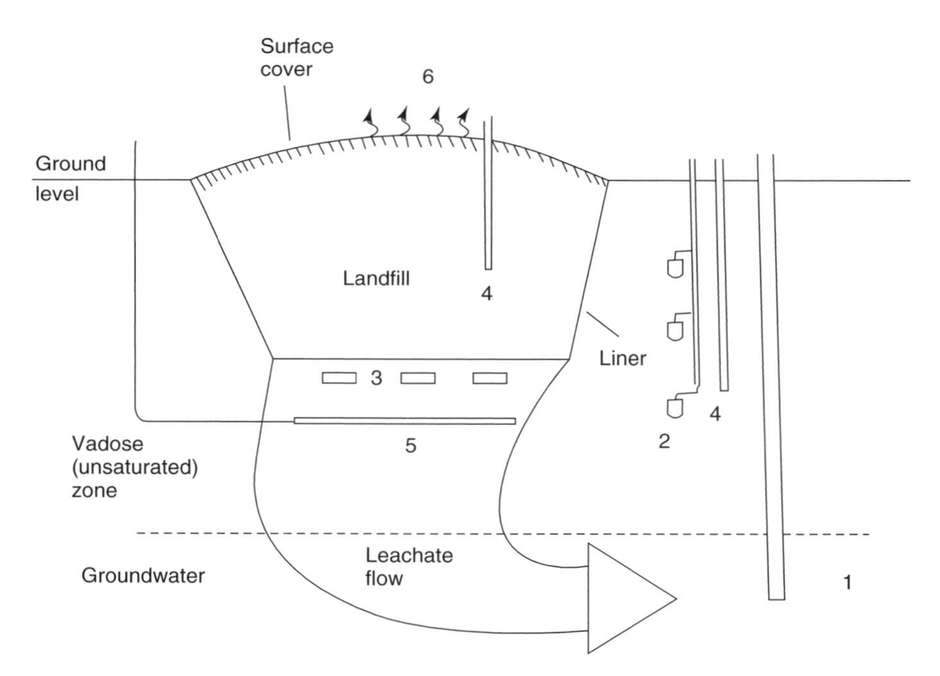

Figure 5.6 Schematic of a typical containment site showing possible sampling methods and locations; 1, water monitoring wells at various depths; 2, suction lysimeters; 3, collection lysimeters; 4, gas monitoring wells; 5, gas sampling probe; 6, surface gas monitoring.

Any escape of liquid from the site will percolate downwards until it reaches groundwater. During design of the site, a preliminary survey should have been made to determine the likely direction of flow and monitoring positions are based on this information. Groundwater is monitored by boreholes sunk to a number of depths. Additional monitoring should be made closer to the site (in particular, underneath the landfill) where the soil has not become saturated with water. This is known as the *unsaturated region* or *vadose zone*. Liquid sampling in this region uses instruments known as *lysimeters*. The most common type for waste site monitoring is the suction lysimeter. This extracts the liquid from the soil by negative pressure inside a porous sampling vessel (Figure 5.7). A more simple design, i.e. a collection (or pan) lysimeter, can be used underneath the containment site. This has the form of a large horizontal tray filled with stone or gravel and covered with a fabric screen. The soil moisture percolates into the tray and subsequently drains into a sump which is accessible from the surface. The liquid is periodically extracted from the sump for analysis.

Gases are sampled both within the ground and on the surface. Ground sampling uses fixed boreholes or sampling probes. These probes are galvanized pipes which

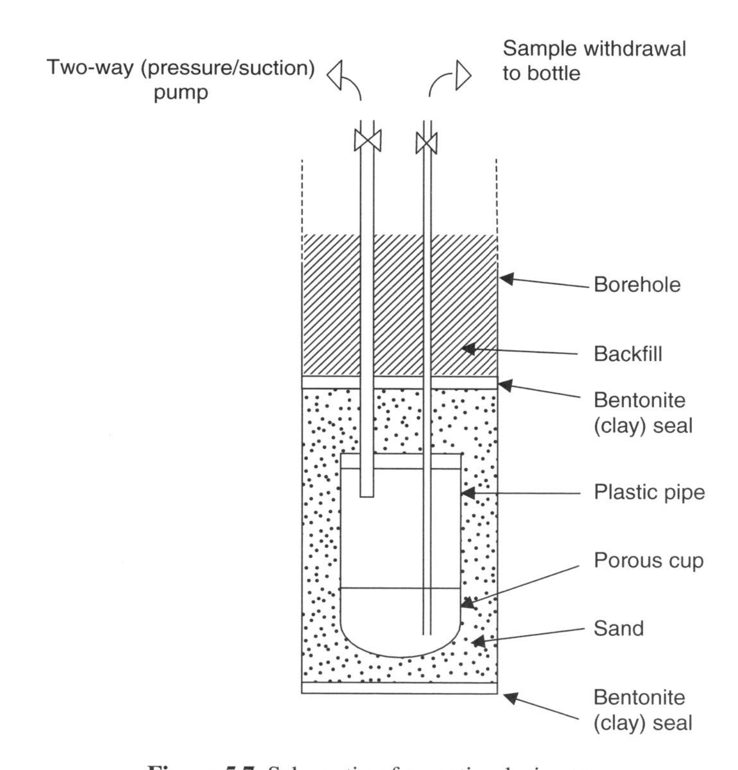

Figure 5.7 Schematic of a suction lysimeter.

can be very easily driven into the solid material and which contain perforations or slots at the required sampling depths. Once again, samples would need to be taken at several depths to gain the overall picture. If the sample is not analysed directly by portable equipment, collection of the vapour could be by any of the standard techniques (adsorbent tube, sample bag or container, or syringe) which will be discussed later in Chapter 6.

Solid samples can be readily putrescible or reactive. They should be stored at 4°C. Water samples should be preserved according to the principles already discussed (see the earlier Sections 3.2, 4.2 and 4.3).

5.6.3 Pretreatment of Solids and Liquids with a High Solid Content

5.6.3.1 Bulk Testing

There are a number of tests which can be performed on solid waste to determine its bulk properties. Chemical procedures include determination of the ash and moisture content and its elemental composition. Prior to testing, the waste may need to be dried, sieved, blended or crushed, and then subdivided to produce a representative sample.

5.6.3.2 Semi-volatile Organics

An extraction stage is necessary prior to GC analysis. The procedure adopted depends on the percentage solid content, as follows:

(i) For <1% solids, liquid–liquid extraction may be used (see Section 4.2.2 above), typically with dichloromethane as the solvent.

(ii) For 1–30% solids, the sample is diluted to 1% with water prior to the extraction as above.

(iii) For >30% solids, sodium sulfate is added to the sample. Extraction can then proceed with a 1:1 acetone/dichloromethane solvent mixture, using ultrasonic agitation.

5.6.3.3 Volatile Organics

For low concentrations (< 1 mg kg^{-1}), the heated solid is mixed with water and the organic is extracted by a purge-and-trap method (see Section 4.2.2 earlier). For high-concentration samples (>1 mg kg^{-1}), an attempt is made to dissolve the compound with methanol. If this is unsuccessful, the solid is mixed with tetraglyme or poly(ethylene glycol). In both cases, water is then added and the extraction proceeds by using a purge-and-trap technique.

For the initial testing of waste, the USA Environmental Protection Agency (EPA) *Toxicity Characteristic Leaching Procedure* exposes the sample to an acetone/sodium hydroxide leaching solution for 18 h. Volatile material is then extracted by purge-and-trap techniques and determined by gas chromatography. If any of a set of volatile organics are present at levels above a threshold value, then the waste is considered to be toxic.

5.6.3.4 Metals

Digestion is typically by refluxing with 1:1 nitric acid with the later addition of hydrogen peroxide and, for some metals (e.g. copper and iron), concentrated hydrochloric acid. After dilution and filtration, or centrifugation, the sample is then ready for atomic spectrometric analysis.

X-ray fluorescence spectrometry (see Sections 5.5.1 and 7.4.1) may also be used to advantage, particularly when employing portable spectrometers, utilizing its capability of elemental analysis regardless of whether the sample is solid, liquid or mixed.

5.6.3.5 Quality Assurance

Quality assurance is of major importance in waste analysis and you may find in standard procedures that blanks are added at several stages of the analytical process (e.g. field, trip and equipment blanks) to check for lack of contamination. In addition, quality checks would be made with spiked samples and spiked blanks (see Section 2.9 earlier).

5.6.4 Analysis of Leachate

The major components and their typical concentrations for a leachate are presented in Table 5.3. You should note the high BOD and COD values (which decrease as the site matures) and also the high concentrations of common inorganic ions.

Trace components include organics from the original waste and also their inter-reaction and breakdown products. You could almost say the leachate could contain virtually any low-molecular-mass organic compound!

Table 5.3 Typical concentration ranges of selected major components in a leachate

Component	Concentration range (mg l^{-1})	
	New site	Mature site
BOD	2000–30 000	100–200
COD	3000–60 000	80–160
Organic N	10–800	80–120
Nitrate	5–40	5–10
Orthophosphate	4–80	4–8
Total hardness	300–10 000	200–500
Chloride	200–3000	100–200
Sulfate	50–1000	20–50
Iron	50–1200	20–200

DQ 5.13

By looking back at Chapter 4, devise a simple preliminary separation scheme which separate the organics into groups of similar compounds.

Answer

You could separate the organics into groups which are extractable under neutral, acidic or basic conditions (see Section 4.2.2). Head-space techniques could be used to separate volatiles. You should have noted from SAQ 4.1 that the USA Environmental Protection Agency uses this type of categorization to classify priority pollutants. Much of the organic material may, however, be non-extractable, partly consisting of high-molecular-mass, and perhaps colloidal, compounds (see Section 3.1 earlier) from the partial decomposition of the organic material in the waste.

Analyses for the extractable components follow the methods discussed previously in Chapters 3 and 4, after taking into account the higher concentrations and more complex analytical matrix involved. The discussion questions given below are to encourage you to examine how these analytical procedures are used.

DQ 5.14

What modifications would you suggest would be needed for the BOD and COD methods described earlier in Section 3.3?

Answer

The BOD values shown above in Table 5.3 are extremely high and the samples would need to be diluted by up to a 1000 times. After dilution, the samples may have low nutrient levels, and thus additional nutrients would have to be considered. The microbial activity necessary for the test may also be inhibited by other components in the waste. For the COD analysis, note the high relative concentration of chloride ions (Table 5.3), particularly in the mature leachate. Mercury (II) sulfate may need to be included in the procedure in order to minimize any potential interference.

5.6.4.1 Trace Organics

In this case, analysis is most often carried out by GC. HPLC with UV detection can be used for the analysis of trace components in leachates which are liable to contain high concentrations of hydrocarbons (e.g. from dumped fuel). The hydrocarbons themselves exhibit no response. You could compare this to analysis by GC (with flame ionization detection) where the chromatogram would be swamped with hydrocarbon peaks unless there was a substantial clean-up stage. A second use of HPLC would be to investigate high-molecular-mass components which are not sufficiently volatile for direct GC analysis. Ion chromatography may be used for non-extractable organic acids.

DQ 5.15

Which GC column type would you think best for determining trace organics in leachates?

Answer

Capillary columns are needed to separate the large number of expected components (see Section 4.2.3 above). Wide-bore, rather than narrow-bore, capillaries would be preferred as these are less likely to be affected by high-molecular-mass impurities. They may be necessary if sample introduction is by a purge-and-trap device.

5.6.4.2 Target Compounds

Preliminary investigations may be simplified if compounds can be identified which can act as *markers* for pollution. These are known as 'target compounds'. A typical application would be to identify if groundwater was polluted. Often, volatile non-polar organics are used.

DQ 5.16

Why do you think *volatile non-polar* organics are used?

Answer

Volatile compounds tend to be relatively small molecules. These are more soluble than their higher-molecular-mass analogues (see Section 2.3 above) and so are more likely to have migrated away from the source. The volatility of the compounds is not a problem in groundwater as there is little chance of vaporization. Highly polar compounds (e.g. acetone, ethanol and acetic acid) can be analysed by GC but they are more problematic than neutral compounds. Such compounds may produce tailing GC peaks or need prior derivitization before analysis.

Marker compounds can also be used to identify compound classes within complex mixtures. BTEX compounds (benzene–toluene–ethylbenzene–xylene(s)) may be used to indicate the presence of petroleum products.

5.6.4.3 Trace Metal Analysis

We have seen earlier in Section 4.3 that metal analysis in natural waters is now largely carried out by using atomic spectrometric techniques. This is also the case for leachate analysis. If spectrometric techniques were attempted, the additional metal ions present in the complex mixture could lead to interferences in the analyses. Leachate sample preparation may be different from what we found with relatively pure natural waters samples. A decision first needs to be made as to whether analysis of dissolved or suspended metal is required. The metal content in the suspended solids may, in fact, be greater than in solution and so any dissolution stage would lead to an unrepresentative analytical concentration. It may be considered better to analyse the solid and dissolved components separately. The sample must be analysed as quickly as possible since the standard preservation technique for metals (acidification) could alter the relative proportions of the dissolved and undissolved phases. If atomic absorption spectrometry is being used, a background correction is necessary due to the complex and largely unknown matrix. Quantification should be by using the standard addition procedure.

5.6.5 Introduction to Gaseous Emissions

The major components and their change in relative concentrations with time are shown in Figure 5.8. Carbon dioxide is the main product when there is a plentiful

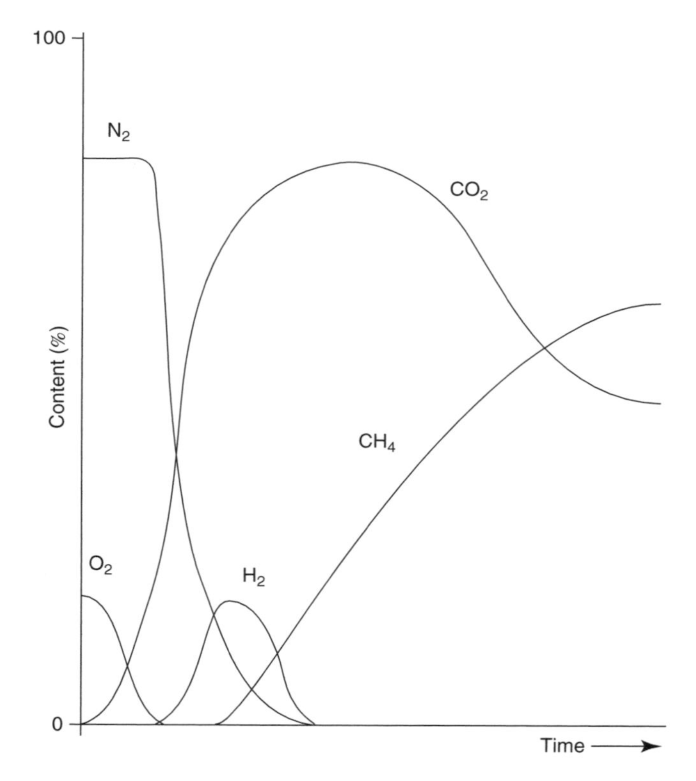

Figure 5.8 The main components of landfill gas emissions and their changes in relative concentrations with time.

supply of oxygen. After a short period of time, the oxygen becomes depleted and degradation continues under anaerobic conditions to produce methane. This continues for many years until all the biodegradable material has been consumed. Some landfill sites collect the methane for use as an energy source. Other components include volatile organics. Individual compounds can reach ppm (v/v) concentrations. These are greenhouse gases and contribute to the volatile organic compound (VOC) loading in the atmosphere. Unpleasant smells and local nuisance can also be produced from the volatile organics, particularly those containing sulfur, and also hydrogen sulfide. The total concentrations of sulfur-containing compounds can reach to ca. 1% under reducing conditions.

Gas analysis is both of the ambient air and sub-surface atmospheres for major and trace components. Details of gas analysis procedures will be discussed later in Chapter 6.

SAQ 5.5

Why do you think care in quality assurance and control is particularly important when analysing waste samples?

5.7 Specific Considerations for the Analysis of Sediments and Sewage Sludge

5.7.1 Sampling and Storage

The first problem with sediment analysis is to *obtain* the sample from the river or sea beds. Core samplers are available for shallow areas. A simple core sampler used for this purpose is shown in Figure 5.9(a). Using this device, a cylindrical tube is first driven into the sediment. On withdrawal, the valve system closes which allows the sample to be withdrawn from the sediment. Just before breaking the surface of the water, the tube is then sealed to preserve the sediment structure. In this way, sections corresponding to different depths in the sediment can be analysed, which can provide a historical record of the deposition of pollutants. Grab samplers (Figure 5.9(b)) may be used for greater depths, or where the sediment is loose so that there is no vertical structure. Dredging can be used for coarse sediments. Samples are often stored deep-frozen.

5.7.2 Pretreatment

DQ 5.17

What is the main difference between sediment samples and other samples we have looked at, which may modify the pretreatment?

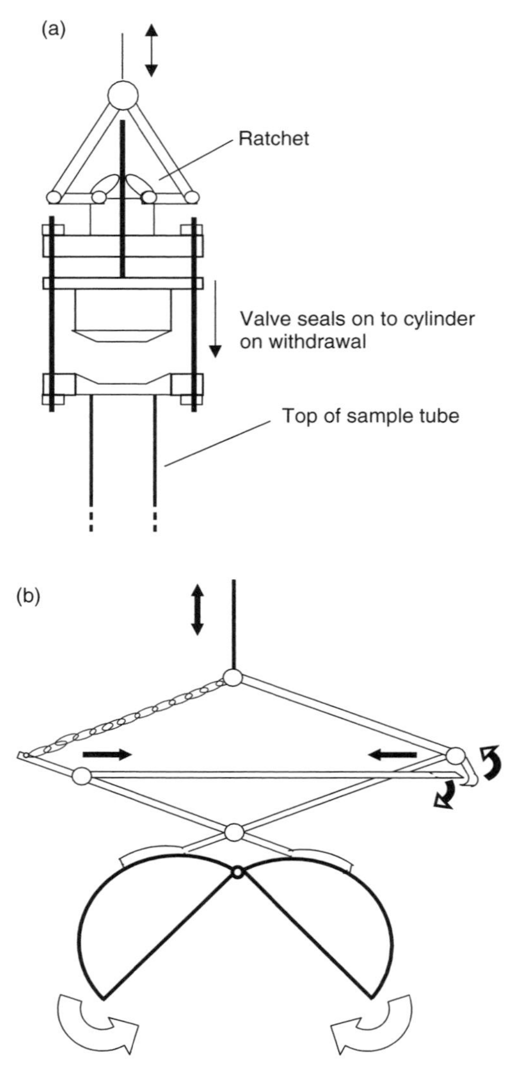

Figure 5.9 Schematics of sediment samplers: (a) simple core sampler (b) grab sampler.

Answer

Quite simply, it is the high water content.

On return to the laboratory the sample is thawed and screened to remove large contaminants such as stones and twigs, if necessary by using pressure. Most of the previous methods we have looked at then dry the sample. It would be quite impracticable to remove the water from sediment by air-drying at room temperature and if the analysis is to determine organic material, the analysis proceeds by using wet samples. Samples for metal ion analysis may be oven dried at 110°C before further treatment. Pretreatment may also include the separation of the sample into size fractions by wet sieving.

DQ 5.18

A map showing the distribution of trace metals around a marine site for the dumping of solid waste showed a wider geographic distribution of metals in the lower size fractions. What do you think might be happening here?

Answer

The smaller particles are drifting more in the local currents.

Can you remember the other reasons why an analysis of size fractions may be useful? Look again at Section 5.1 above if you are unable to do this.

5.7.3 Extraction Techniques for Organic Contaminants

Organic analysis is once again based on solvent extraction of the homogenized slurry (produced by using a high-speed blender), often using a Soxhlet apparatus. The extracting solvents we have previously discussed are often extremely hydrophobic. If they were used with a wet sediment sample, we would end up with water–solvent emulsions. In order to overcome this, more polar extractants such as acetonitrile or acetone are used. Concentrations of organic pollutants are often in the $\mu g\ kg^{-1}$ range.

5.7.4 Dissolution Techniques for Trace Metals

Acid dissolution is used for the analysis of **adsorbed** metal ions but care has to be taken not to dissolve the bulk sediment itself. A suitable acid mixture would be concentrated nitric acid and hydrogen peroxide, with metal concentrations in the $mg\ kg^{-1}$ region being expected.

Extreme solubilization techniques have to be used for analysis of the less-soluble portion of the sediment. A typical method uses hydrogen fluoride under pressure in a 'Teflon'-lined 'Parr' bomb. We will be looking later at other methods for insoluble solid analysis in Chapter 7.

DQ 5.19

Why do you consider that in some cases analysis of adsorbed metal ions is most relevant for a study of environment problems, whereas in other cases analysis of the whole of the sediment is more appropriate?

Answer

Environmental analysis is concerned with ions and compounds which are available to living species. Only loosely adsorbed metal ions will ***always*** *be readily available. The availability of the ions in the bulk of the sediment will depend, among other criteria, on the particle and the chemical composition.*

5.7.5 Analysis of Sewage Sludge

Sewage sludge is included in this section due to the high water content of such samples, which require similar pretreatment for trace-level organics to those discussed above for sediment samples. The material, however, has a high organic content and so digestion is necessary before any metal analysis is carried out. Typically, this would involve heating with concentrated nitric acid in a Kjeldahl apparatus (see Figure 5.3) and extraction of the metal ions after dilution with water.

SAQ 5.6

What are the relative merits of investigating metal or insoluble organic compound pollution in a river or sea by:

(a) analysis of the water;
(b) analysis of the sediment;
(c) analysis of seaweed
(d) analysis of the fish or shellfish?

5.8 New Extraction and Dissolution Techniques

We have seen that the standard method for extraction of low-volatility organics from solids is by Soxhlet extraction.

DQ 5.20

Why do you think that other techniques are being considered as alternatives to this well-established and thoroughly tested technique?

Answer

A Soxhlet extraction takes several hours to perform. During that time, one piece of apparatus is dedicated to a single extraction. Large quantities of solvent are needed, typically 300 ml per extraction, and there is no automation. As so far described, it would not easily fit into a high-throughput laboratory.

Similar conclusions concerning the lack of automation could be made for metal-ion digestion. Over the past few years, a number of techniques have been developed to overcome these problems and are gradually finding acceptance within the standard methods. All are commercially available. These techniques are described separately in the following sections and then a comparison is made between them.

5.8.1 Automated Soxhlet

Instruments are available from a number of manufacturers which attempt to overcome the difficulties discussed above, while still retaining the Soxhlet principle of extraction into fresh solvent. Typically, the instruments can digest four to six samples simultaneously, using only approximately a fifth of the solvent in a conventional Soxhlet and at five times the extraction rate. The increased speed is achieved by using a two-stage extraction arrangement, with the first of these having the sample directly in the heated solvent. It is only after this preliminary step that the thimble is raised above the solvent and is washed by refluxing solvent in the conventional Soxhlet manner. Some systems then allow controlled evaporation of the solvent rather than needing a separate concentration stage.

5.8.2 Accelerated Solvent Extraction

This is an automated technique where the extraction of the organic from the solid can take minutes rather than the hours required for Soxhlet extraction. This is achieved by extracting at elevated temperatures (typically 100°C) while maintaining a pressure (1500–2000 psi) to prevent the solvent from boiling. Conventional solvents (or mixtures), such as dichloromethane, perchloroethylene or hexane/acetone, are used and so previously developed methods can be transferred with little modification. Automatically weighed samples (5–15 g) are fed into a sealed container which is then heated. After the predetermined extraction time, the extract is automatically removed to a second container, together with fresh solvent which has been used to flush the extraction vessel. The extract can then be concentrated for the later analytical stages. Typically, an extraction takes 15 min to complete and uses 15 ml of solvent. Commercial equipment can process, sequentially, batches of 24 samples (perhaps consisting of 20 samples, a spiked sample, a spiked sample duplicate, a blank and a standard), and the technique has been accepted in at least one EPA standard methodology. Methods are available for semi-volatile pollutants and for volatiles such as BTEX compounds.

5.8.3 Microwave Digestion and Microwave-Assisted Extraction

An alternative method by which digestion or extraction can be speeded up is to introduce energy to heat the sample in the form of microwaves. Instruments are available from a number of manufacturers, some of which allow introduction of sample and reagents into the extraction vessels by using flow systems, while others employ manual operation. A typical microwave oven could process six to twelve samples simultaneously. The extraction vessels are glass or 'Teflon' and the sample can be blanketed with clean air or nitrogen. The 'Teflon' vessels may be pressurized and operate at temperatures up to 250–300°C. Accurate temperature control is necessary for thermally labile compounds.

One prerequisite of using microwaves is that you need a compound capable of absorbing the radiation in the extraction vessel. If water is present in the sample or in the extraction solvent, there is no problem since this is an excellent microwave absorber. If insufficient water is present, care needs to be taken that the extraction solvent or solvent mixture can absorb the radiation. This is achieved by having at least one highly polar component in the system.

The technique can be used for acid digestion of solids for subsequent metal analysis and is accepted for several EPA methods. The water in the acid is the microwave absorber. Typically, the digestion of a 0.5 g sample takes 30 min to carry out.

When applied to the extraction of organics, the technique is known as Microwave-Assisted Extraction. Typical analyses would be for PAHs and total petroleum hydrocarbon (TPH). Extraction times are less than 30 min and use around 30 ml of solvent per extraction. If a non-polar extraction solvent is used, then you would have to ensure that there was sufficient residual water in the sample to absorb the microwave radiation. Polar-solvent mixtures can also be used, such as hexane–acetone, dichloromethane–acetone or methanol–toluene.

5.8.4 Sonication

In this method, a vessel containing the sample and extraction solvent is immersed into an ultrasonic bath. Heat may also be applied. The procedure could include a period of continuous sonication, followed by a period of intermittent sonication, perhaps a few minutes every hour for several hours. Alternatively, a succession of batch extracts (say, 10 min sonication for each) could be made, combining the extracts before the following concentration stage. The advantages are seen as the use of less complex laboratory equipment and the possibility of many simultaneous extractions in one apparatus. The extraction time can be less than that of a typical Soxhlet extraction. This simple technique is now part of standard methods for extraction of semi-volatiles from waste with a high solids content (see Section 5.6.3 above).

5.8.5 Supercritical Fluid Extraction

A *supercritical fluid* is a substance which is maintained above a critical temperature and pressure where there is a single fluid phase rather than distinct gas and liquid phases. Below this critical point (temperature/pressure), the compound

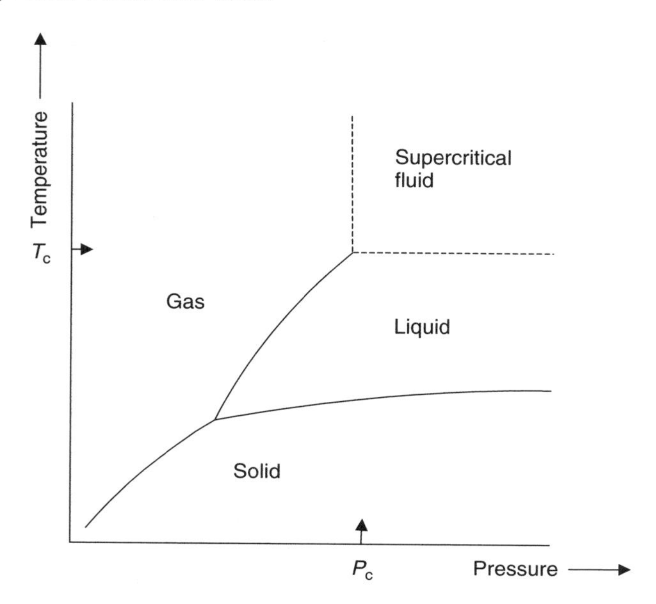

Figure 5.10 A phase diagram for a typical gas, illustrating the formation of a supercritical fluid above the critical (temperature/pressure) point: T_c, critical temperature; P_c, critical pressure.

reverts to being a gas or a liquid. This behaviour is illustrated by the phase diagram shown in Figure 5.10.

The properties of a supercritical fluid can be seen as being midway between a gas and a liquid. It resembles a liquid in that substances have a limited solubility in the fluid, while it resembles a gas by having a low viscosity and surface tension. The low viscosity allows the fluid to penetrate porous solids easily, and hence the current interest for solid extraction. A number of gases may be used to produce the supercritical fluid but most analytical applications now use carbon dioxide. The latter has a low cost and low toxicity. Carbon dioxide has a critical temperature of 31°C, which means extraction can take place at only slightly elevated temperatures. Its critical pressure is 74.8 atm. Typical extraction conditions would be 50°C and 400 atm, with an extraction time of ca. 30 min. You should note that this temperature is lower than those used in other extraction techniques, which may be an advantage for thermally labile compounds.

Once the compound of interest has been extracted, the solvent can simply be removed by lowering the pressure. There are a number of possibilities available for the collection of the extract. Volatile compounds can be transferred directly to a gas chromatograph or can be adsorbed on to a solid for subsequent desorption into the GC system. Semi-volatile material can be collected either as a dry solid or in an appropriate liquid solvent.

Carbon dioxide is a non-polar molecule and is an ideal solvent for non-polar compounds, although the solubilities of even moderately polar compounds can be low. Solubilities can be improved by the addition of a few percent of modifiers such as methanol or acetone which increase the polarity of the solvent. Temperature and pressure also affect the solubilities. A second advantage of supercritical fluid extraction is that the extraction process can be made highly selective by changes of modifier, temperature and pressure. The extraction efficiency can sometimes, however, be highly matrix-dependent.

Applications of the technique are now appearing in some standard methods, although development of the more widespread use of the technique may be slowed by the unfamiliarity of analysts to the sometimes unexpected properties of supercritical fluids.

5.8.6 Comparison of the Techniques

Previous comparison of techniques throughout the book have been made on the basis of ease of use, cost, and applicability to small or high-throughput laboratories. A further consideration with extraction methods is the efficiency of analyte extraction. This, of course, will be dependent on the analyte and sample matrix, as well as with the solvents and conditions used for the extraction. A number of publications are listed in the Bibliography at the end of this text which attempt such comparisons and lead to the obvious conclusion that the optimum method is analyte- and matrix-dependent. The choice of the most generally applicable technique may become clearer over the next few years as further practical experience is gained, particularly in the ease of use of the competing technologies in high-throughput laboratories. Soxhlet extraction, however, remains the reference method.

SAQ 5.7

New extraction techniques from solids are often compared by spiking a clean sample with the analyte and then determining the extraction efficiency. For a given sample, many techniques produce a reproducible extraction efficiency which can be considerably less than 100%. The extraction efficiency has been found in some cases to decrease with increase in the time-span between addition of the spike and the analysis. Comment on these facts with respect to the environmental effects of contamination.

Summary

Most chemical analytical techniques rely on the analyte being present in solution. This chapter examines extraction and dissolution techniques from solids to solubilize the components of interest. The analysis can then proceed by the instrumental techniques which have already been discussed in earlier chapters. Solids which are of importance in studying the environment include animal and plant specimens, soils, contaminated land and waste and landfill sites, sediments and sewage sludge, and atmospheric particulates. Specific extraction and dissolution procedures have been discussed for each type of solid, except for atmospheric particulates. In some instances (particularly for landfill sites), the sampling and analysis of associated liquids and gases is also described. Atmospheric particulates are dealt with later in Chapter 7, after discussion of the gaseous components of the atmosphere in the next chapter.

Chapter 6
Atmospheric Analysis – Gases

Learning Objectives

- To be able to list the major components in external atmospheres and appreciate the need for analytical monitoring.
- To understand the difference in type and concentration of pollutants in external and internal environments and the difference in approach needed for their analysis.
- To realize the importance of personal sampling.
- To be able to describe, compare and contrast the analytical methods available for external and internal atmospheres.
- To understand simple methods of flue gas analysis.
- To appreciate the availability of portable instruments for gas analysis and the possibilities of remote sensing.

6.1 Introduction

From your previous knowledge and from the earlier chapters in this present text you should have some idea of the major components of the atmosphere.

DQ 6.1

List the components of clean dry air and give an indication of their approximate concentrations.

Answer

These components and their typical concentrations are shown in Figure 6.1 below.

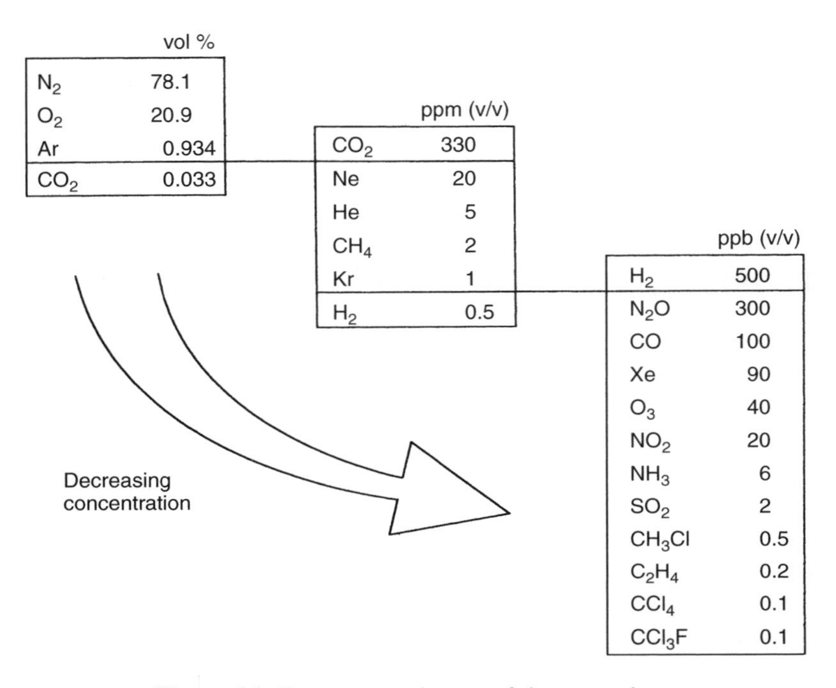

Figure 6.1 Gaseous constituents of the atmosphere.

You will probably have included the major components, but many of the minor components might be a surprise to you. You may have considered some of these to be anthropogenic pollutants. There are few of the common gases (chlorofluorocarbons being possibly the only example) which are not found in the atmosphere from natural sources. Non-localized atmospheric pollution problems are mainly concerned with increases in concentrations of naturally occurring compounds above the unpolluted clean-air levels.

We have previously mentioned in Chapter 1 the problems due to acid rain which contains high concentrations of sulfur and nitrogen oxides. The compounds are oxidized over periods of hours or days to sulfuric and nitric acids by reactions which may include other atmospheric components (e.g. ozone and particulates). Global warming was also discussed with increasing carbon dioxide levels being a major contributing factor. The problem arises from the increased absorption of infrared radiation by carbon dioxide. Introduction of other covalent compounds into the atmosphere can add to the problem, particularly if they absorb at wavelengths which would otherwise not be absorbed by the atmosphere (*window regions*). This can occur with compounds containing C–H, C–Cl or C–Br bonds.

Concern over specific compounds includes the rising concentrations of methane (another greenhouse gas) and of ground-level ozone (an oxidant, which produces breathing difficulties) which is increasing even in non-industrialized areas. Localized problems in urban or industrial areas can be more complex, not only due to the introduction of a large number of other pollutants but also from atmospheric reactions producing new species. A good example of this is the complex series of reactions which occur each day in large cities in hot climates throughout the world. Under specific meteorological conditions (thermal inversion where layers of hot light air are found above cold dense air, producing a stable atmospheric condition), pollutants build up in the atmosphere without dispersal. Gases given off by vehicles (CO, NO, NO_2, unburnt hydrocarbons, etc.) inter-react to produce a range of oxidants including ozone and peroxyacetyl nitrate (PAN). The chemical changes that occur throughout a typical day are shown in Figure 6.2. These reactions produce a haze over a city, known as photochemical smog, and the compounds produced can cause respiratory problems. There is a more general concern over the emission of all volatile organic compounds (VOCs) into the atmosphere. Many are toxic in their own right, all are greenhouse gases, and they may contribute to the atmospheric chemical reactions discussed above.

One of the main reasons behind current environmental concern is the potential effect of pollutants (including airborne pollutants), either directly or indirectly,

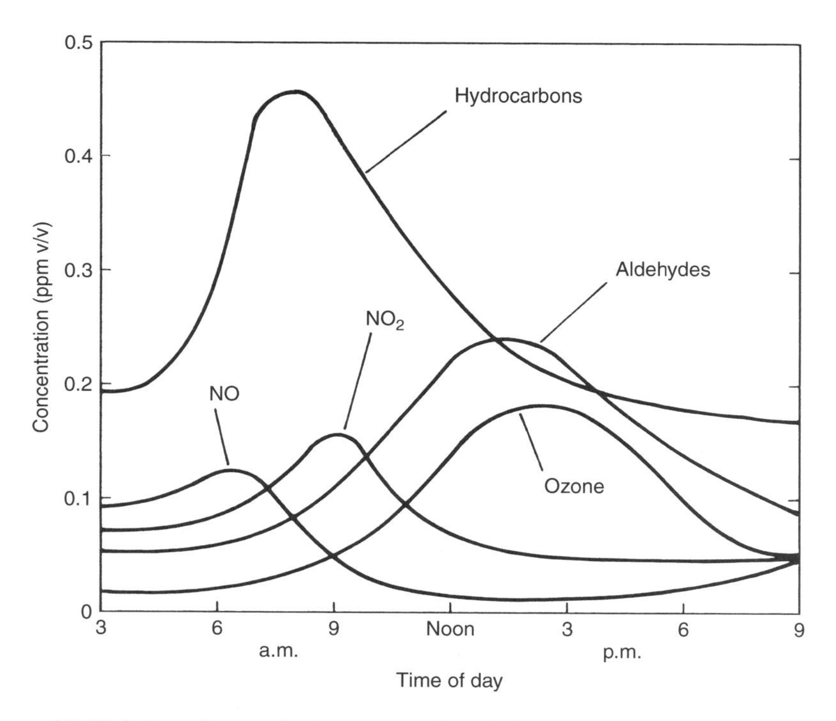

Figure 6.2 Variation of atmospheric concentrations during a photochemical smog incident.

on human health. Most of the population of the industrialized world spend their days inside buildings at work or at home. The monitoring of atmospheres within buildings (internal atmospheres) is then also of major importance. Internal atmospheres are enclosed, thus preventing dispersal of any pollutants. Higher concentrations of pollutant gases than are found in external atmospheres may be expected.

Internal atmospheres can also have a much wider range of pollutants than are found in external atmospheres. Large numbers of potentially hazardous chemicals are produced or used inside buildings, for example:

- Gases from fuel combustion
- Solvents in paints
- Gases from cleaning fluids

There are many unexpected sources. Even some forms of wall insulation can give off a hazardous gas (i.e. formaldehyde).

Measurement of ambient concentrations is, however, just one of the many types of gas analysis which you may have to perform. Emission concentrations (concentrations of pollutants in flue gases or vehicle exhausts before they are dispersed into the environment) are just as important, with legislation often being based on these discharge values. Compare this with water monitoring where legislation and necessary analysis concerns both discharges and the receiving water. Emissions from diffuse sources such as waste disposal sites or reclaimed land will need monitoring. Some examples of these are shown in Figure 6.3.

Exhausts and flue gases can contain a much wider range of gaseous compounds than are likely to be detectable in the general atmosphere. Concentrations in the wider atmosphere may not build up to detectable levels because of the continuous removal of the gases by physical and chemical processes. Look at the list of gases which may be emitted from a typical coal-fired power station, as shown in Figure 6.4, and notice the discharge of hydrogen chloride, hydrogen fluoride and even mercury.

You may not be personally involved but analysis of the upper atmosphere is important when trying to understand the effect of pollutants discharged on the earth's surface upon the ozone layer.

Concentrations of pollutants in atmospheres and exhaust streams can vary significantly over a short time-period and, for many purposes when monitoring atmospheres, the average concentration over a period of time is required as well as an instantaneous measurement. These are known as Time-Weighted Average (TWA) concentrations. In the next two sections, you will find that some of the analytical methods are very suited to the measurement of time-weighted concentrations, as the analyte is collected over an extended time-period. One reading can then be used to give the average concentration over the whole sampling period. These methods are described below in Section 6.2. Other methods, which are used mainly to give instantaneous concentration readings, are described later in

Vehicle exhausts
Urban atmospheres

Flue gases
External workplace atmospheres

Internal domestic
atmospheres

Reclaimed-land
emissions

Personal monitoring
Internal workplace atmospheres

Figure 6.3 Some examples of local monitoring requirements.

Section 6.3. In order to use these methods for TWA measurements, many readings are necessary, but this can nowadays often be readily achieved by microprocessor control and data storage.

What, though, are the most useful time-periods over which to take an average reading? The concentration of many components in an external atmosphere varies over a 24 h cycle. Diurnal influences include the effect of sunlight and factory emissions. A concentration averaged over the 24 h cycle may be appropriate. The USA National Air Quality Standard (NAQS) for sulfur dioxide (365 $\mu g\ m^{-3}$ or

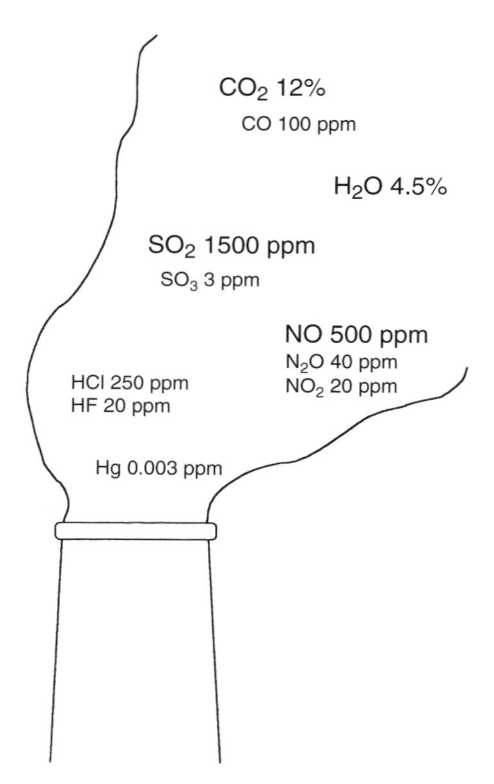

Figure 6.4 Typical flue-gas analysis of a coal-fired power station.

140 ppb, averaged over a 24 h period) is an example. Shorter-term averages may be required for pollutants which vary rapidly throughout the day. The USA standards (NAQS) for carbon monoxide are based on 8 and 1 h periods.

Sampling times of 24 h would not be appropriate for internal atmospheres. The major concern for pollutant levels in internal atmospheres is over human health and, in particular, chronic exposure. The sampling time is then generally specified to be over an 8 h period, thus reflecting the length of the average working day.

You may come across a number of terms related to time-averaged exposures which are derived from national legislation. For example, the United Kingdom specifies an Occupational Exposure Standard (OES) for most gases (Table 6.1) which is defined as the concentration of an airborne substance, averaged over the reference period, at which there is no evidence that it is likely to be injurious to persons exposed by inhalation. More toxic substances, such as benzene and hydrogen cyanide, are assigned a Maximum Exposure Limit (MEL), which is defined as the maximum time-averaged concentration to which a person can be lawfully exposed by inhalation.

Table 6.1 Typical 8 h average Occupational Exposure Standards (OESs)[a]

Gas/vapour	Concentration	
	ppm (v/v)	mg m^{-3}
Ammonia	25	18
Carbon monoxide	30	35
Methanol	200	266
Nitrobenzene	1	5

[a]UK, 2000.

Although the reference period is normally 8 h, there are also short-term (15 min) standards which apply to any period throughout the working day. This is to accommodate the possibility of acute effects from the gas. The 15 min OES for ammonia is 35 ppm or 25 mg m^{-3}.

6.1.1 A Note on Units

The concentrations given in Table 6.1 are expressed as **volume** of analyte/total **volume** of sample. This is a common way of expressing gas concentrations, with many direct-reading instruments calibrated in these units. We perceive gas concentrations as volume fractions, rather than in other units. Most people would know whether an atmosphere containing 20.9 vol% oxygen would support life, but would they be so sure if it contained 9.3×10^{-3} mol l^{-1} oxygen or 299×10^3 mg m^{-3} oxygen? There is, however, one difficulty. The same units cannot be used for measuring suspended particulates. As we will see later in Chapter 7, these are just as much of concern in the environment. The alternative units based on analyte **weight**/total **volume** may be used for both gases and particulate material, with typical concentrations being expressed as follows:

- μg m^{-3} for external atmospheres
- mg m^{-3} for internal atmospheres

Gas concentrations throughout the remainder of this book will be given in both of these units. You will find concentrations expressed as weight/volume measurements in most national and international legislation. For everyday purposes, volume/volume measurements are also frequently used, due to the convenient form of their expression.

The conversion between ppm and mg m^{-3} is straightforward, simply requiring the relative molecular mass of the compound, and the molar volume of the gas (for an approximate conversion this can be taken to be 24.0 l for all gases at 20°C and 1 atm pressure). However, the units are often quoted side-by-side in tables of environmental standards. Thus, the UK 8 h occupational exposure standard for toluene is expressed both as 50 ppm and 191 mg m^{-3}.

DQ 6.2

The current EC monthly limit for nitrogen oxide emissions from coal-fired power stations (measured as NO_2) is 650 mg m^{-3}. What is this concentration in parts per million (volume/volume)?

Answer

The relative molecular mass of nitrogen dioxide = 46
Therefore, the number of moles of nitrogen dioxide in 1 m^3 air

$$= \frac{650 \times 10^{-3}}{46}$$
$$= 14.1 \times 10^{-3} \text{ mol}$$

The volume occupied by 1 mole at 20°C and one atmosphere pressure

$$= 24.0 \text{ l} = 0.0240 \text{ m}^3$$

Therefore, the volume of nitrogen oxide in 1 m^3 air

$$= 14.1 \times 10^{-3} \times 0.0240$$
$$= 338 \times 10^{-6} \text{ m}^3$$

Therefore, the concentration of nitrogen oxide = 338 ppm volume/volume.

Look back at Table 6.1 for other examples of the two sets of units. By using my approximate molar volume for all gases, the conversion can be expressed as follows:

$$\text{concentration (ppm)} = \frac{\text{concentration (mg m}^{-3}) \times 24.0}{\text{relative molecular mass}} \tag{6.1}$$

The expression is identical if you convert μg m^{-3} to ppb. For compounds with molecular masses close to 24 (e.g. ammonia and carbon monoxide), concentrations expressed as μg m^{-3} and ppb are numerically roughly the same, whereas for higher-molecular-mass compounds (e.g. nitrogen dioxide and nitrobenzene) the numerical value of the weight/volume concentration is higher than the volume/volume concentration.

SAQ 6.1

What is the difference in meaning in the term 'parts per million' when applied to *gas* concentrations and *aqueous* concentrations?

SAQ 6.2

Briefly summarize the expected concentration ranges of pollutants in external and internal atmospheres, and in exhaust gases.
Suggest reasons why we may find there are sometimes different analytical methods used for external and internal atmospheres.

6.2 Determination of Time-Weighted Average Concentrations

6.2.1 Absorption Trains

This is perhaps the method which would first occur to you for monitoring trace components. A known volume of gas is bubbled through an absorbing solution. At the end of the sampling period, the solution is taken back to the laboratory for analysis, generally using volumetric or spectrometric methods. Ion chromatography may also be used.

The absorption train consists of a number of containers through which the gas sample is drawn. The sample volume is measured by a gas meter, but for shorter sampling times where the flow can be kept constant, the gas flow and an accurate sampling time may be used instead. An International Organization for Standardization (ISO) typical specification is shown in Figure 6.5. A typical practical system for monitoring atmospheres would be as shown in Figure 6.6. Individual components of the train can vary according to the specific needs of the analysis.

The reagents used in the Drechsel bottle(s) (see Figure 6.6) are determined by the gas to be analysed. There are specific reagents for most of the inorganic gases, including SO_2, Cl_2, H_2S and NH_3. Carbon monoxide is the only common exception. A typical procedure is shown by the West and Gaeke method for sulfur dioxide, which uses a spectrometric final analysis. The sulfur dioxide is absorbed in an aqueous solution of sodium tetrachloromercurate, and the colour developed by the addition of *p*-rosaniline hydrochloride (in hydrochloric acid) and formaldehyde. The absorbance is measured at 560 nm:

$$H_2O + SO_2 + HgCl_4{}^{2-} \longrightarrow HgCl_2SO_3{}^{2-} + 2H^+ + 2Cl^- \qquad (6.2)$$

The above procedure is now a reference method for the USA Environmental Protection Agency (EPA), i.e. it is judged to be the best available technique and can be used to assess other methods.

DQ 6.3

What properties must the absorbent possess in order to produce an accurate analysis?

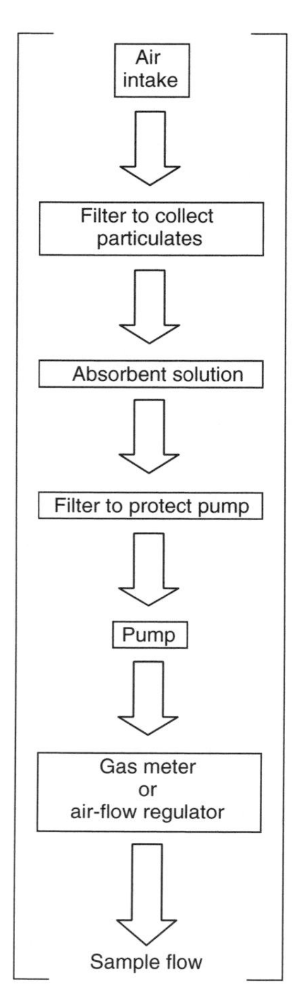

Figure 6.5 Schematic of a typical absorption train.

Answer

1. *The reagents have to be highly specific to the analyte gas.*
2. *The absorption of the analyte has to be quantitative. Remember that you may be analysing compounds whose concentration may only be parts per billion (v/v) in the atmosphere.*
3. *The reagent has to be resistant to oxidation and to being stripped from solution. Remember that you are bubbling air through the solution for periods of up to 24 h.*

Figure 6.6 shows a sampling train monitoring the external atmosphere and conveniently located in a building. This ideal situation may not always be possible

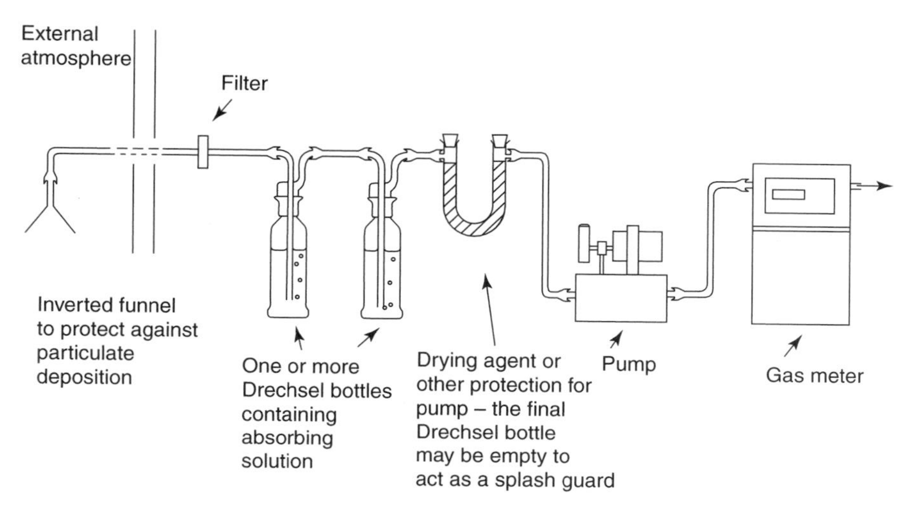

Figure 6.6 Schematic showing the typical components of a gas absorption train.

to achieve and the train may be located outside, but under shelter. Apparatus is available with the complete train enclosed in a single, portable container. External sampling sites are often determined by security considerations. Rooftops of municipal buildings are often used, even though an environmentally more appropriate position may be street level.

6.2.1.1 Flue Gas Analysis

With a few modifications from that shown in Figure 6.6, absorption trains can be used for flue gas analysis. The concentration of the gas may vary both across the flue and along its length. Preliminary practical and/or theoretical work is necessary to determine the optimum sampling location(s). The gas may contain highly corrosive components (see Figure 6.4) and so the sampling tube is 316 grade stainless steel or a higher grade of corrosion-resistant alloy. If the train is being used for gas sampling alone (the train may also be used to sample particulates), there should be a glass wool plug in the line to filter the gas. The gas will certainly be at an elevated temperature and will probably be saturated with water. If you are unsure where the water is coming from, try performing a simple calculation to determine the mass of water produced by combustion of 1000 g of a typical alkane such as hexane. Condensation in the sampling tube may occur if the temperature is allowed to fall before reaching the absorbent solution. To prevent this, the sampling tube should be heated. The train is after that point similar to the one shown in Figure 6.6 – one or more Drechsel bottles (which may be in ice–water to prevent evaporation), and then a drying column, followed by a pump and gas meter. You should note that the gas-phase concentrations of many of these phase components are determined by temperature dependent equilibria, for example:

$$NO + O_2 \rightleftharpoons 2NO_2 \tag{6.3}$$

$$SO_2 + O_2 \rightleftharpoons 2SO_3 \tag{6.4}$$

In order to obtain the true high temperature concentration, any cooling process in the sample train has to be rapid so as to freeze the equilibria at the flue temperature values.

There are many advantages in using absorption trains for gas analysis. We will come across some other applications below in Sections 6.3.1 and 7.2.5. At the end of this main section, you will be asked to compare the methods with alternative procedures.

6.2.2 Solid Adsorbents

The most commonly used method for low-concentration volatile organic compounds, particularly for internal atmospheres, is to adsorb the gas on to a solid and later analyse the components by gas chromatography.

6.2.2.1 Sampling

Passive and active sampling methods can be used. Passive samplers (sometimes called diffusion samplers) consist of the adsorbent (typically activated charcoal or 'Tenax' porous polymer) contained in a small tube sealed at one end, with the other end being exposed to the atmosphere. The adsorbent is separated from the atmosphere by a diffusion zone which is either an air gap, or an inert porous polymer, according to the manufacturer. The tubes may be clipped to the lapel or carried in the breast pocket to allow personal monitoring.

An alternative design of sampler is in the form of a badge which can also be clipped to the lapel. The principle of operation is similar to the tube design described above. Examples of the two types are shown in Figure 6.7.

Active sampling methods draw air through the sample tube by means of a pump. Sampling rates can be of the order of several hundred ml min^{-1}, although lower flow rates are often used (as low as 20 ml min^{-1}) so that sampling can continue over an 8 h period without the capacity of the tube being exceeded. Some adsorption tubes (Figure 6.8) contain two sections of adsorbent, the main section to be used for the analysis, while the second (back-up) section is used as confirmation that the capacity of the analytical section has not been exceeded.

Pumps are available which are small enough to be clipped to the waist with the sample tube positioned on the lapel (as shown in Figure 6.9.) We will come across these 'personal samplers' again in the following chapter on particulate analysis. The advantage of active sampling over passive sampling is that lower concentrations can be monitored for a given sampling time.

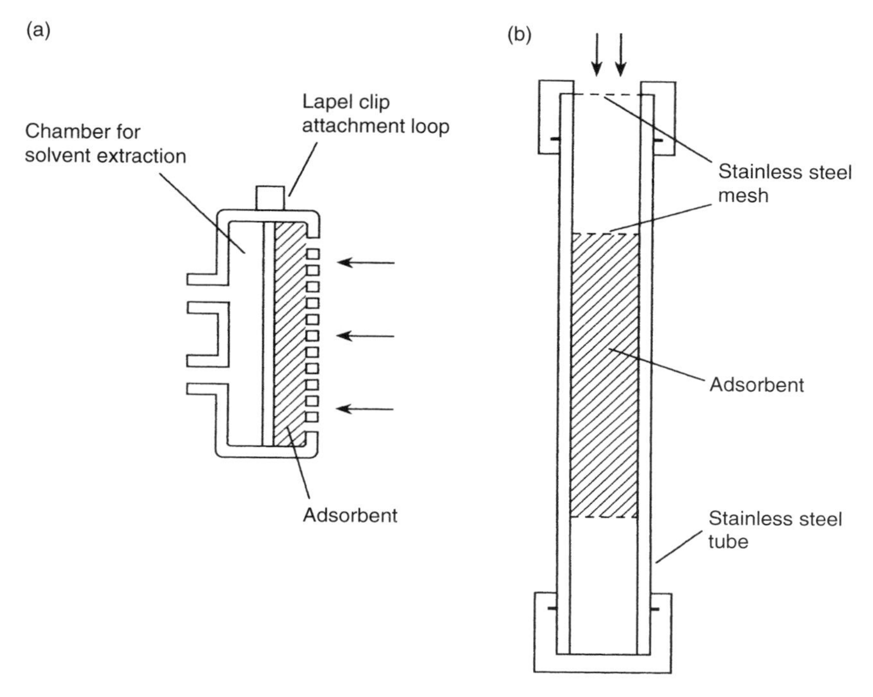

Figure 6.7 Examples of passive (diffusion) samplers: (a) badge type; (b) tube type.

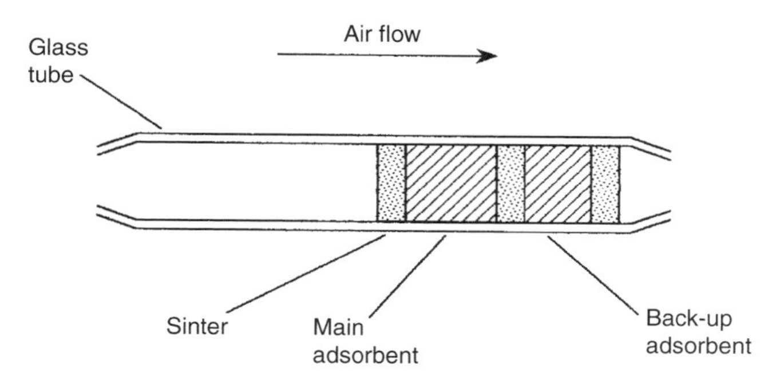

Figure 6.8 Schematic of a typical adsorption tube used for active sampling.

6.2.2.2 Desorption of Sample

Transfer of the analyte to the chromatograph is either by thermal desorption or by solvent extraction. The thermal desorption method involves similar equipment to that used in the analysis of aqueous organic compounds employing purge-and-trap techniques (see Section 4.2.2 above). Solvent extraction requires mixing the

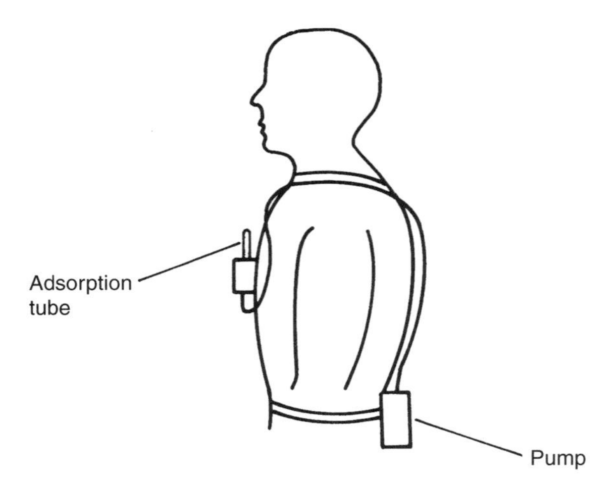

Figure 6.9 Illustration of personal sampling.

adsorbent with a fixed volume of solvent to extract the analyte, followed by injection of the extract into the gas chromatograph.

DQ 6.4

What problems can you foresee in the use of these sampling methods in quantitative analysis?

Answer

The major problem is in the absorption and desorption efficiencies of the sampling. *Although, with standard methods, adsorption can be assumed to be 100%, desorption may be less than this and will be different for each compound. Standard analytical methods (e.g. the UK 'Methods for the Determination of Hazardous Substances' (MDHS) series) recommend specific adsorbents to use for each analyte, but even so, the desorption efficiency has to be measured for each new batch of tubes and has to be included in the analytical calculations.*

A second problem is the possibility of overloading the adsorbent. *The theoretical capacity of the adsorbent (known as the 'breakthrough volume') can be found, either from published tables, or by experimental determination, by passing a gas of known composition through the tube and monitoring the effluent air with a flame ionization or similar detector. This volume is, however, influenced by many factors, including the presence of water vapour, other organic compounds and temperature, and so will be different for each analysis.*

6.2.2.3 Chromatographic Analysis

The chromatographic separation is usually straightforward when using standard columns. There is usually little problem with regards to detector sensitivity for monitoring internal atmospheres and flame ionization detection is often used. The only difficulty is in the choice of extraction solvent. Solvents which do not show a response with a flame ionization detector (e.g. carbon disulfide, which is toxic and has a low flash point) are often hazardous materials in their own right! For routine analyses, quantification is by comparison of peak areas with standard solutions injected into the chromatograph after correcting for the desorption efficiency.

The latter is ideally calculated by analysis of a standard gas mixture, although this is not always practicable. A less rigorous approach is to inject a known quantity of pure compound on to the adsorbant and to measure its recovery on extraction.

6.2.2.4 Production of Standard Gas Mixtures

There are various methods available, dependent on the particular requirements, for producing standard gas mixtures. These include the following:

(i) Small volumes (up to a few litres) of reference gas can be produced by injecting a known volume of the pure compound, as a liquid, through a septum into an enclosed volume of gas and allowing the liquid to vaporize.

(ii) If a continuous flow of reference gas is needed, then dynamic methods are necessary. Permeation tubes are often used for this purpose. These contain the volatile organic compound within a small PTFE tube which allows slow permeation of the vapour through its walls into a known flow of gas. The rate of diffusion is adjusted by changing the temperature of the tube over the range from ambient to 40°C. The concentration produced in the gas stream can be calculated from the weight loss of the permeation tube over a given time-period and the gas flow rate.

(iii) An alternative dynamic technique for gases or volatile liquids is to inject the compound into a gas stream at a constant rate by using a syringe-pump.

6.2.3 Diffusion (or Palmes) Tubes

The majority of investigations which determine time-weighted average concentrations use one of the two methods described above. There have, however, been a number of applications of the use of diffusion tubes (also known as Palmes tubes) over the past few years, particularly where a large number of sites are being simultaneously monitored. The method incorporates features of the two techniques already described but has the advantage of simple and easy-to-construct samplers which can readily be taken to and left at the sampling site.

The unit, as shown in Figure 6.10, consists of a short tube (standard dimensions are 7.1 cm length, and 0.95 cm inside diameter) which is open at one

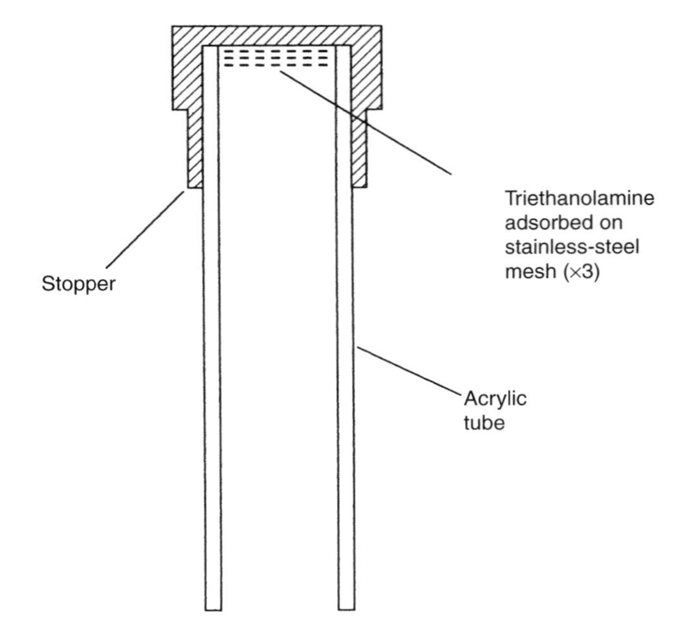

Figure 6.10 Schematic of a diffusion tube.

end and has a liquid adsorbed on to stainless steel mesh at the closed end of the tube. The method relies on the natural diffusion of the gas into the liquid.

The reagent is typically exposed to the atmosphere for several weeks, after which the absorbed gas can be determined by standard analytical techniques (e.g. spectrometry or ion chromatography). The principle of the technique is that the rate of absorption is determined by the rate of diffusion of the gas along the tube. *Fick's law* states that the rate of diffusion of a gas is proportional to the concentration gradient. The concentration at the open end of the tube is the ambient concentration. At the closed end of the tube, this is assumed to be zero as it is being continuously absorbed by the liquid. Hence, the rate of diffusion is proportional to the atmospheric concentration.

The original validation of the technique was for internal atmospheres where air currents (which could possibly affect the rate of diffusion and hence the accuracy and precision) would be low. The technique, however, has now been successfully applied to external atmospheres and has been used in a number of major atmospheric investigations in the UK. The most common use of the technique so far has been for the determination of nitrogen dioxide. The absorbent liquid is triethanolamine, and the analysis is completed by spectrometric analysis (at 550 nm) of the nitrate released, using sulfanilamide and *N*-(1-naphthyl)ethylenediamine hydrochloride.

For a sampling tube of the stated dimensions operating at 21°C, we can write:

$$\text{Concentration of NO}_2\text{ (ppb)} = \frac{Q_{NO_2} \times 1000}{2.3 \times \text{exposure time (h)}} \qquad (6.5)$$

where Q_{NO_2} is the quantity of nitrogen dioxide absorbed (nmol).

The factor '2.3' is determined from the known value for the diffusion coefficient of the gas and the tube dimensions. The precision of the technique is not large (the variance was found to be 10% under ideal conditions). This can be partially compensated for by the low cost of the apparatus, thus allowing groups of 10 or more tubes to be left at each sampling position. Other applications include the analysis of NH_3, SO_2, O_3 and BTEX compounds.

SAQ 6.3

For routine monitoring of sulfur dioxide in external atmospheres by using an absorption train, aqueous hydrogen peroxide is often used as an absorbent, rather than the West and Gaeke reagent (see above):

$$SO_2 + H_2O_2 \longrightarrow H_2SO_4$$

What are the advantages and disadvantages of hydrogen peroxide for large-scale monitoring exercises?

SAQ 6.4

Compare and contrast active and passive sampling for monitoring internal atmospheres.

SAQ 6.5

Alternative methods in the 'Methods for the Determination of Hazardous Substances' series (UK) for toluene in atmospheres use solvent extraction and thermal desorption techniques prior to GC analysis.
What do you see as their relative merits?

6.3 Determination of Instantaneous Concentrations

6.3.1 Direct-Reading Instruments

Instruments are available to monitor individual gases over the whole range of concentrations we have been discussing. We will start by looking at instruments designed to be transportable to the monitoring site for measuring ambient concentrations and then later discuss methods for workplace and personal monitoring.

This section finishes with techniques which may be used for monitoring atmospheres directly, without the need for sampling.

Instruments for atmospheric ambient monitoring are often based on spectrometric techniques (chemiluminescence, infrared, fluorescence, etc.).

DQ 6.5

Which of these techniques are potentially the most sensitive and so most suitable for the low concentrations found in ambient air?

Answer

The techniques involving light emission (chemiluminescence and fluorescence) are potentially the most sensitive.

6.3.1.1 Chemiluminescence and Fluorescence

The chemiluminescent method used for nitrogen oxides is based on the following reactions:

$$NO + O_3 \longrightarrow NO_2{}^* + O_2 \quad (6.6)$$

$$NO_2{}^* \longrightarrow NO_2 + h\nu \quad (6.7)$$

where $\lambda = 600\text{–}875$ nm.

Ozone, generated within the instrument, is mixed with the sample under reduced pressure and the light emission monitored with a photomultiplier, thus giving a measurement of the nitric oxide concentration of the sample (Figure 6.11). Total nitrogen oxides can be analysed by thermal conversion of nitrogen dioxide to nitric oxide before analysis. The nitrogen dioxide concentration is then calculated by difference from the two readings. This is the EPA reference method for NO_2 and is also the specified method in EU legislation. The detection limits are approximately 10 ppb (18 $\mu g\ m^{-3}$).

The same reaction can also be used to monitor atmospheric ozone. A second chemiluminescent method can also be used, based on the reaction of ozone with ethylene and monitoring the light emission at 430 nm. This method has the advantage of little interference from the presence of NO. The detection limits are approximately 1 ppb (2 $\mu g\ m^{-3}$).

Sulfur dioxide can be measured, without chemical pretreatment, by gas-phase fluorescence spectrometry, giving a limit of detection of 2 ppb (5 $\mu g\ m^{-3}$).

DQ 6.6

What method for the production of calibration gases, which has already been discussed, could be incorporated into these instruments?

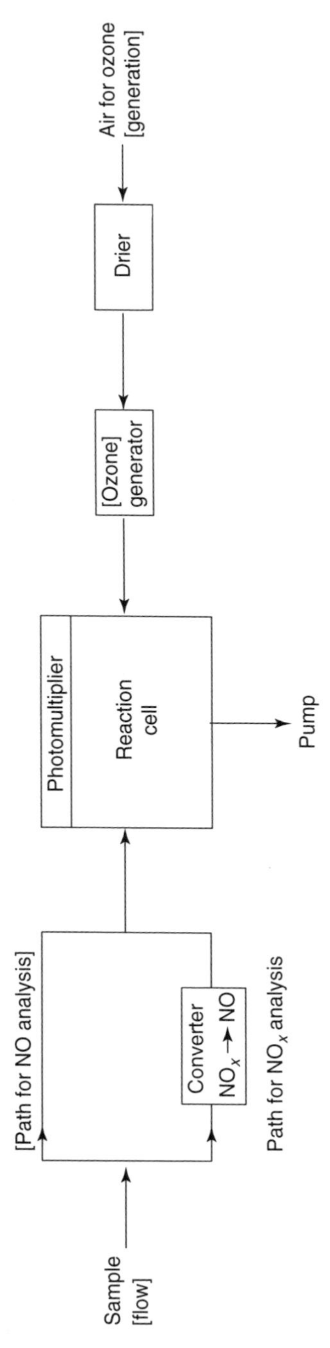

Figure 6.11 Schematic of a chemiluminescent analyser for nitrogen oxides.

Answer

Calibration methods often use standard gas mixtures generated by using permeation tubes (see Section 6.2.2 above).

6.3.1.2 Infrared Spectrometry

Infrared absorption spectrometry is commonly used in workplaces for monitoring a wide variety of inorganic gases and organic vapours.

DQ 6.7

What molecules are capable of absorbing infrared radiation?

Answer

All molecules which contain more than one element, i.e. in physical chemistry terms, all molecules except monoatomic and homonuclear diatomic molecules. From an environmental point of view, this includes all molecules except O_2, N_2, He, Ar and the other noble gases.

The spectra can be highly complex and each molecule gives a unique absorption pattern (Figure 6.12).

As in other areas of the electromagnetic spectrum, the Beer–Lambert law applies for the absorption of radiation. If you look back at the mathematical formulation of the law (Section 3.4.1) you will find the absorbance of radiation is proportional to the molar concentration of the absorbing species and cell pathlength. In order to maximize the sensitivity for low-concentration gases, long pathlengths need to be used. This can be achieved by having large sample cells ($\leq$ 1 m) and also reflecting the radiation many times through the cell, thus giving a total pathlength of up to 50 m. Instruments relying on absorption spectrometry for gas analysis tend to be bulky, even though they may still be classed as 'portable'.

These instruments may be of similar design to the complex spectrometers which you will be familiar with in the laboratory, which often measure the absorption of radiation after separating the infrared radiation into its spectral components. These are termed *dispersive* infrared spectrometers. By monitoring the absorption at different wavelengths, a large number of gases can be analysed by a single instrument. One manufacturer pre-programmes the instrument to be able to analyse more than 100 gases and vapours. An alternative design is sometimes used where no spectral separation is necessary. These are known as *non-dispersive* spectrometers (Figure 6.13). You may wish to attempt to deduce how the instrument operates before reading the next paragraph.

When a molecule in a gas absorbs infrared radiation, the net effect is to heat the gas. It can only absorb radiation at the frequencies which are specific to the molecule. The gas present in the detector cell will heat up and expand only if radiation of a suitable wavelength enters the cell. There is no impediment for

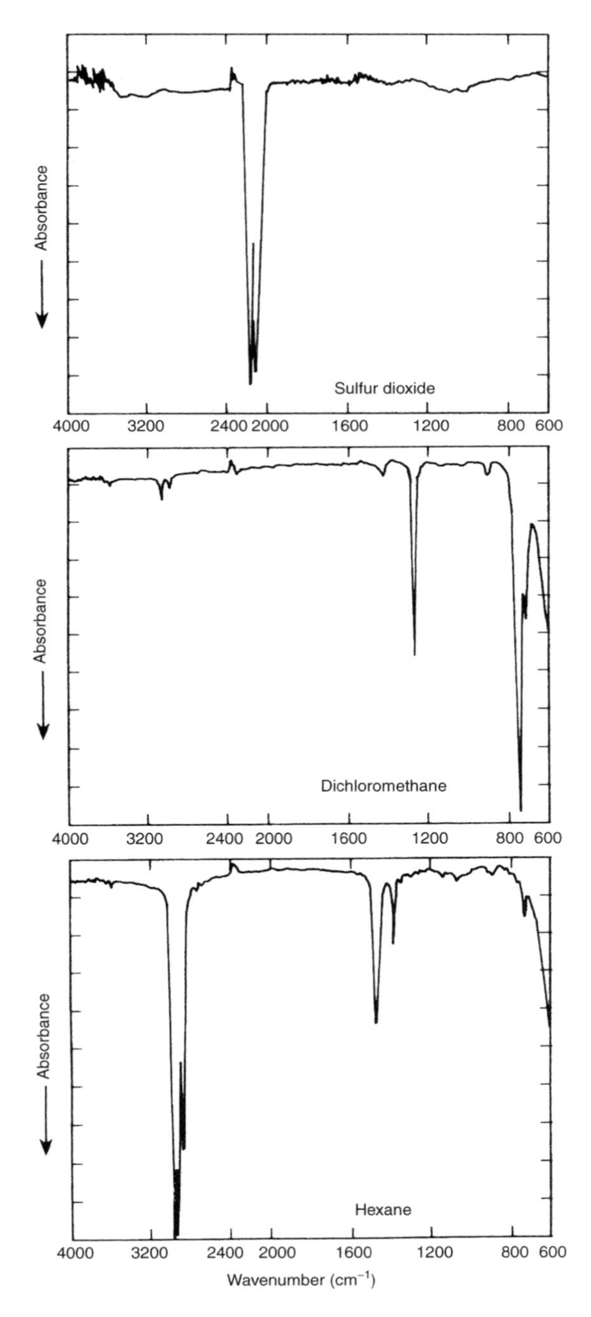

Figure 6.12 Some typical infrared spectra of volatile and gaseous compounds.

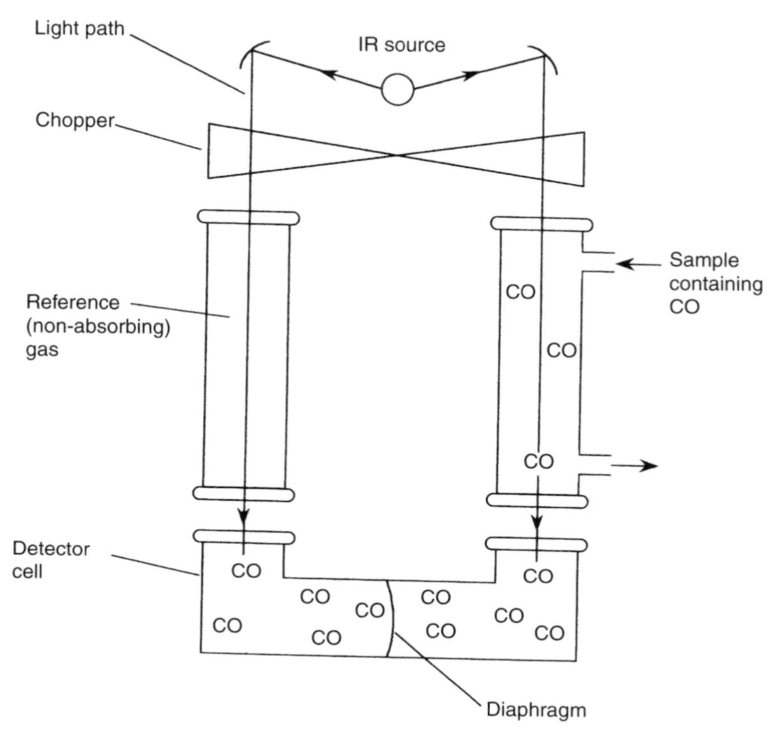

Figure 6.13 Schematic of a non-dispersive infrared carbon monoxide analyser.

suitable radiation on the left-hand side of the cell to reach the detector, and the gas will heat up in the detector cell. On the right-hand side, some of the radiation at the carbon-monoxide-specific wavelengths will already have been absorbed by the carbon monoxide in the sample. The heating of the detector cell will be lowered. The net effect is to distort the diaphragm towards the right-hand side. When the beams are turned off, the diaphragm will return to its original shape. By chopping the beams, an oscillation will be produced. The size of this oscillation will measure the concentration of the carbon monoxide in the gas.

Non-dispersive instruments are available for a number of gases, including carbon monoxide (EPA reference method), carbon dioxide, sulfur dioxide, acetylene, methane and water vapour. Although each is sold as a separate instrument, remember that the only major difference is the gas within the detector cell. It is this gas which gives the instrument its specificity.

DQ 6.8

Infrared spectrometry is a very suitable method for analysis of gases when there are few absorbing species but there may be difficulties when there are many species. Why do you think this is the case? What alternative method do you think would be more suitable?

Answer

The complex nature of infrared spectra means that there is the possibility of overlap in absorptions in multi-component mixtures. A chromatographic method (see Sections 6.2.2 and 6.3.3), which separates the components before quantification would be more suitable.

6.3.1.3 Application of Spectrometric Methods to Flue Gas Analysis

The most obvious method would be to place a spectrometer in a sampling train such as that described in Section 6.2.1 above. The spectrometer would operate at ambient temperature and so the flue gas would have to be cooled before reaching the instrument. However, there are a number of potential problems with this approach. As the gas temperature falls, solids may condense and so a filter is necessary immediately prior to the spectrometer. There may also be condensation of water. The latter problem can be overcome in a number of ways. If the sensitivity of the instrument is sufficient, the sample gas can be diluted with clean dry air to prevent the condensation. Alternatively, a permeation drier can be used in which the gas passes through tubes constructed of 'Nafion', a synthetic polymer which is selectively permeable to water. A chiller can also be used, rapidly cooling the sample to 3–5°C to remove the water (Figure 6.14).

DQ 6.9

What property of a gas would make the cooling and water-condensation approach unsuitable? Give an example of a gas with this property.

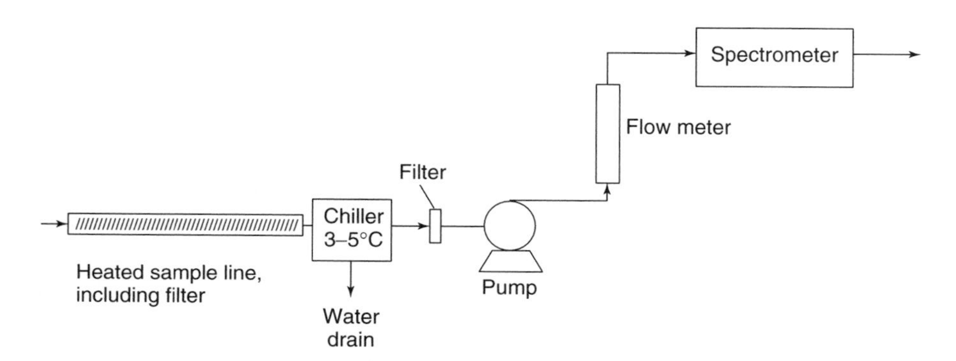

Figure 6.14 Schematic of a typical sampling train for the spectrometric determination of flue gases.

Answer

A high solubility in water. The most problematic of the common gases in this respect is hydrogen chloride.

For hydrogen chloride and other water-soluble gases, an alternative method would be to use a spectrometer which can operate at flue-gas temperatures. Hydrogen chloride is analysed by this method using infrared spectrometry. Chemiluminescence and UV spectrometers are also available to measure NO_x and sulfur dioxide, respectively. You should, however, bear in mind when using these instruments that the concentration measured is for the gas stream containing moisture. Recalculation of the concentrations to dry values (i.e. removing the moisture contribution) is necessary to provide data comparable with room-temperature methods.

Infrared spectrometry may be used to determine the concentrations in flue gases directly by placing an IR source and the detector on each side of the flue. Instruments are available for CO, CO_2, SO_2, CH_4 and N_2O. A narrow-wavelength range is chosen which includes the absorption wavelength of the compound of interest and has minimum interference from other gases. There are many problems with this approach, such as the absorption of radiation by dust in the gas and the need to keep the optical components clean. One solution, as used in a carbon monoxide analyser, is to have a double beam of radiation, i.e. an analytical beam and a reference beam. Both beams will be equally affected by the unwanted absorptions. You will almost certainly have already come across double-beam UV or IR spectrometers which are used in the laboratory because they have better stability to background and radiation source variations than single-beam instruments. After both beams pass through the flue and before the detector, one beam passes through a cell containing pure carbon monoxide. The residual beam after the cell will be unaffected by small variations in carbon monoxide in the flue gas and acts as the reference. Optical components are kept clean by purge streams of clean air, although the instrument still requires recalibration every few minutes by inserting cells containing pure gas into the analytical beam.

6.3.1.4 Electrochemical Sensors

As we have already found, there is often a need for personal monitoring within a workplace environment, particularly for industrial gases (e.g. NH_3, CO_2, Cl_2, HCN, HCl, H_2S and SO_2) which may quickly build up in atmospheres through leakages, or accumulate in unventilated areas. Personal monitors are available for individual gases which are based on electrochemical sensing, with a different sensing head being required for each gas. The reaction of the analyte gas at an electrode produces a current which is proportional to its gas-phase concentration. The monitors normally possess a concentration display and also an audible alarm if the pre-set maximum concentration is exceeded. Instruments need to be periodically recalibrated (perhaps every few weeks) for the zero reading and, by using a standard gas mixture, at a value close to the maximum of the measuring range.

The cheapness today of computer storage and data processing is having a major effect on workplace and personal monitoring. The advent of data loggers means that results may be stored for later processing. Some sensors have data loggers as an integral part of the instrument, and thus the exposure of an individual over a working day can be quickly recalled. Such information could have implications for the establishment of safety levels. Current levels are based on either 8 h long-term or 15 min short-term time-weighted averages (see Section 6.1 above). Logging sensors are typically capable of measuring and storing 10 s averages. They can show that, even at a single location, instantaneous measurements vary dramatically and that a simple average may not be a good method of representing the true exposure of an individual. The person may be exposed to far higher levels for short periods of time. This can have severe effects on an individual's long-term health.

DQ 6.10

Portable site instruments use infrared absorption techniques, whereas personal instruments use solid-state electrochemical methods. Can you explain why different methods are used for each application?

Answer

Infrared absorption varies with concentration and pathlength according to the Beer–Lambert law. For maximum sensitivity for low-concentration gases, long pathlengths are necessary (several metres). This leads to bulky instrumentation, even when multiple reflections are achieved in the sample cell. In addition, optical components tend to be heavy, and need careful alignment – very undesirable features for personal instruments.

On the other hand, solid-state electronics are lightweight and rugged, thus making these ideal for personal monitors. Electrochemical techniques, however, do require more frequent calibration than spectrometric methods, so making them less appropriate for long-term background analysis.

6.3.2 Gas Detector Tubes

Hand-held and easy-to-use instruments are often needed for monitoring internal atmospheres where high concentrations of hazardous gases can quickly accumulate. One simple type of instrument uses gas detector tubes.

Such tubes are available from a number of manufacturers for most common inorganic gases and volatile organic compounds. Trade names include 'Draeger', 'Gastec', 'RAE Systems' and 'Kitagawa'. They are constructed of glass, are several centimetres in length, and are packed with an analyte-specific reagent adsorbed on to an inert solid. A fixed volume of gas is drawn through the reaction tube by using a hand pump. This may be a bellows-type or piston pump according to the manufacturer. A typical example is shown in Figure 6.15.

The sample time is a few seconds, during which a colour develops from the sampling end of the tube. At the end of the sampling period, the colour should extend along a fraction of the length of the tube. The tubes are pre-calibrated with a concentration scale on the glass surface so that the distance the colour has travelled can be directly related to the gas concentration.

The colour may be produced by a number of methods. In some cases, a coloured product is formed from colourless reagents. Hydrogen sulfide detection tubes contain a colourless lead salt adsorbed on to silica gel. The product is black lead sulfide, according to the following reaction:

$$Pb^{2+} + H_2S \longrightarrow PbS + 2H^+ \qquad (6.8)$$

In other cases, the colour is due to an indicator change. Detector tubes for carbon dioxide contain hydrazine as the reactant and crystal violet as the redox indicator. The reaction with carbon dioxide causes the indicator to change colour to purple, as follows:

$$CO_2 + N_2H_4 \longrightarrow H_2N{-}NH{-}CO_2H \qquad (6.9)$$

A range of tubes is available for each gas to accommodate different concentration ranges. These ranges are typical of internal atmospheres and emission concentrations (ppm to percentage levels), but can in some cases be extended to lower levels. This is achieved by increasing the gas volume sampled by using a continuous pump rather than a bellows system. Under these circumstances, care has to be taken that the reagent will not be stripped from the support or oxidized (check in Section 6.2.1 above for similar potential problems encountered with

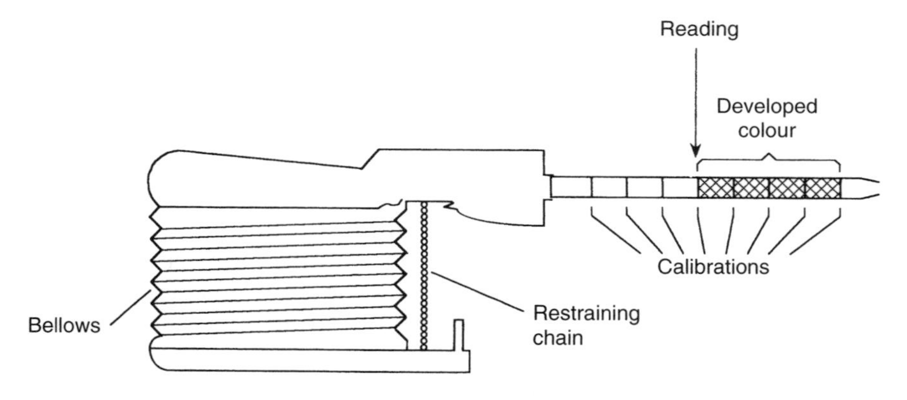

Figure 6.15 Schematic of a gas sampling tube with bellows.

reagents in absorption trains). The tubes are not generally used for sampling periods of longer than a few minutes, and hence are not suitable for very low level contamination.

Two further problems need to be considered before the application of these tubes:

(i) *Precision*. The relative standard deviation varies from compound to compound. In the most favourable cases (e.g. hydrogen sulfide detection), where the chemical reaction proceeds rapidly, the relative standard deviation is 5–10%. In less favourable cases (e.g. mercury detection), a relative standard deviation of 20–30% may be found.

(ii) *Interferences*. These are well-known for inorganic gas detection and specified in manufacturers' literature. Sometimes, a separate zone of reactive solid is included in the detector tube to remove potential common interferences before they reach the calibrated layer. Carbon monoxide tubes contain a zone of chromium (VI) to remove hydrogen sulfide, benzene and other organics. Interferences may, however, be more serious in the detection of specific organic compounds. As an example, adjacent members of the homologous series give positive indications on hexane tubes.

6.3.3 Gas Chromatography and Mass Spectrometry

We will consider first those GC methods where a sample of gas is introduced directly into the chromatograph without preconcentration. This technique finds application for the analysis of inorganic gases (e.g. O_2, N_2, CO and CO_2), particularly in exhausts or flues, and also for gas streams containing mixtures of volatile organic compounds. The chromatograph may be situated in a laboratory, but could also be a portable design which could be carried to the monitoring site, or it could be permanently positioned at the sampling point, away from the laboratory. The following discussion is centred on the use of a laboratory-based instrument, with additional comments relating to any differences in portable or site-based systems.

6.3.3.1 Sampling

A large variety of containers may be used for sampling the gas, as shown in Figure 6.16. Portable instruments usually include a small pump to draw in gas through a sampling tube.

DQ 6.11

What problems do you see in sampling gases and injecting them directly into a GC system?

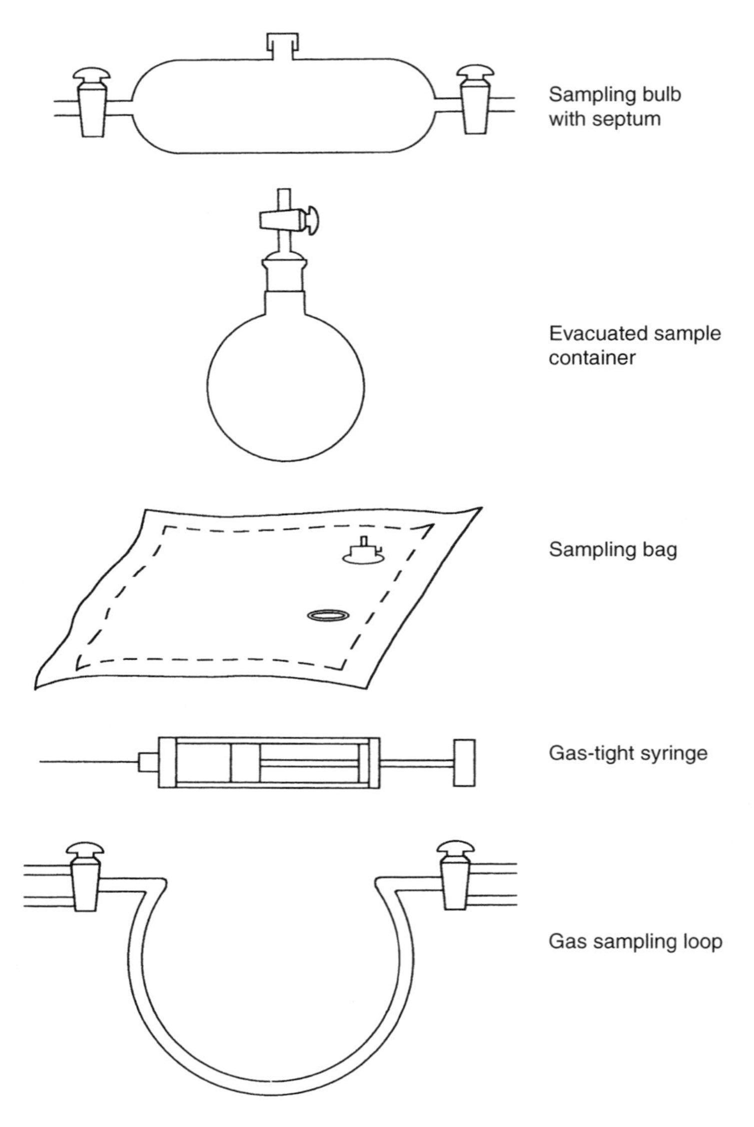

Figure 6.16 Some examples of the equipment used for gas sampling.

Answer

Large sample vessels are necessary, typically several hundred millilitres in size.

It is difficult to check for any leakage/contamination of the sample.

Minor components may be lost by reaction on the walls of the vessel.

Injection of large volumes of gas into the chromatograph disturbs the carrier gas flow.

6.3.3.2 Chromatographic Analysis

Gas–solid chromatography is used for the separation of inorganic gases and low-molecular-mass organic compounds. Molecular sieves are often used for permanent gases (Figure 6.17). These separate gases in order of their molecular size. Unfortunately, one of the most important components of flue gas, i.e. carbon dioxide, is permanently adsorbed by the molecular sieve. A silica gel column, which separates by adsorption, is needed for this gas. Note that the other common inorganic gases are not well separated on this column and so a complete flue gas analysis would require both columns.

Organic porous polymer adsorbents may be used for both low-molecular-mass organic compounds and inorganic gases. Several manufacturers produce a series of stationary phases, e.g. the 'Porapak' series. The chromatographic separation of the gases can be optimized by a suitable choice of stationary phase within the series.

DQ 6.12

Which gas chromatography detector have you come across which is suitable for inorganic gases such as oxygen, nitrogen and carbon dioxide?

Answer

The thermal conductivity detector (Katharometer) is suitable for all gases. Flame ionization detection, which is often said to respond universally, will not easily detect most inorganic gases.

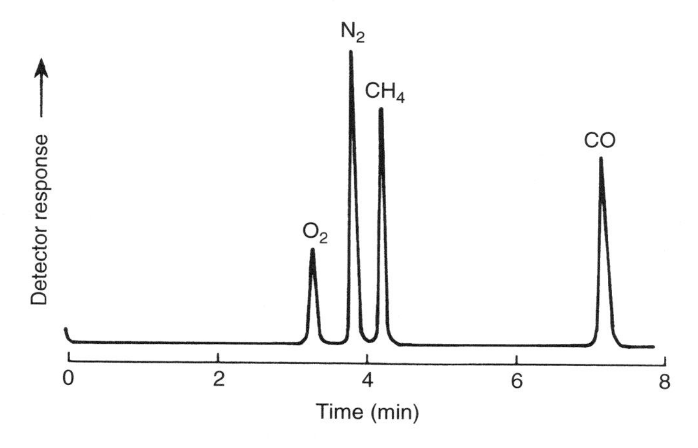

Figure 6.17 Separation of a gas mixture on a 5A molecular sieve column.

The thermal conductivity detector has relatively low sensitivity and so cannot be used for trace analysis, with the lower limits of detection being in the region of a few hundred parts per million. The greatest sensitivity can be achieved by the use of a low-molecular-weight carrier gas. Hydrogen would give the greatest sensitivity, but its use is often discouraged on safety grounds. A helium carrier gas gives slightly lower sensitivity than hydrogen. However, the cost of helium varies enormously worldwide and in some countries it is a very expensive option.

An alternative procedure is possible for carbon monoxide detection where the gas is reduced to methane by using a nickel catalyst. This can then be detected with high sensitivity by using flame ionization detection.

Lower concentrations of volatile organic compounds can be determined by using conventional gas–liquid chromatography after first concentrating the gases using either solid-phase adsorbents (see Section 6.2.2 above), by concentrating the analyte in a liquid nitrogen trap, or by solid-phase microextraction (see Section 4.2.2 earlier).

6.3.3.3 Portable Chromatographs and Mass Spectrometers

DQ 6.13

In what ways do you think a portable gas chromatograph may be different from a laboratory instrument?

Answer

Changes would have to include a reduction of size, weight and the number of gases, plus the utilities used in order to make the chromatograph portable.

Most manufacturers of portable gas chromatographs use a chromatography column which can separate the components at, or near, ambient temperature. This removes the need for a high-temperature oven. A carrier gas is an essential component of a gas chromatograph. Some instruments use cleaned ambient air rather than cylinder gas. Detectors which do not need additional gas supplies are also favoured, although flame ionization is sometimes used for organic analyses. This form of detection has the disadvantage for portable and site instruments in requiring the maintenance of the detector flame but has the advantage that the detection response for hydrocarbons is similar and so a single calibration can be used for multi-component mixtures. Thermal conductivity detection can be used as an alternative for high-concentration components. At least one manufacturer uses photoionization detection for trace organic analysis.

DQ 6.14

Can you think of a method for determining the total organic vapour content of an atmosphere without determining each component separately?

Answer

One method would be to inject a sample directly into a flame ionization detector without passing it through a chromatographic column. The response would then be proportional to the total organic content.

This is the basis of commercial volatile organic (hydrocarbon) content (VOC) monitors. Some portable chromatographs with flame ionization detection have an option to bypass the column in order to measure the total organic vapour concentration.

Although you may think that GC–MS is a specialist laboratory technique, such has been the recent miniaturization and increase in robustness that it can now be found in site installations. We have mentioned above in Chapter 4 that the use of an MS selective detector may mean that prior separation of components can be less rigorous. Perhaps we could have a mass spectrometric gas monitor without the need for a chromatograph at all? Portable quadrupole spectrometers are now available which can determine individual alkanes and also chlorinated and sulfur compounds. Typical applications would be contaminated land, industrial site and flue gas emission monitoring.

6.3.4 Monitoring Networks and Real-Time Monitoring

The current concern over air quality requires co-ordinated monitoring on a large scale, perhaps nationwide or even internationally. This could include the following species:

- Gases leading to acid rain
- Vehicle exhaust gases
- Ozone

Ideally, the monitoring stations would be automatic and this puts a number of constraints on the type of instruments used, as follows:

- The instruments should require as few consumables as possible.
- They should be self-calibrating.
- If possible, they should provide continuous monitoring.
- They should be capable of being connected to a central control and data collection system.
- The equipment should be capable of being stationed at the monitoring location.

A number of small-scale programmes, which had been running over many years in the UK, were rationalized in 1995 into national networks with a current total of 93 urban and 19 rural monitoring sites. The species determined and the techniques used are summarized in Table 6.2.

Automatic calibration is on a daily basis and uses a blank produced from ambient air cleaned by passing over solid adsorbents, plus a calibration gas.

Table 6.2 Techniques used and compounds detected in the UK automatic monitoring network

Compound	Technique	Number of sites
CO	Techniques include non-dispersive IR spectrometry	73
NO_x	Chemiluminescence	87
O_3	UV absorption at 254 nm	65
SO_2	Fluorescence	67
25 Hydrocarbons (including benzene and butadiene)	Gas chromatography after preconcentration on solid adsorbents, followed by thermal desorption into a cold trap	13

The calibrant is produced from either permeation tubes (NO_x and SO_2) or an internal generator (O_3), or it can be a calibration gas mixture (CO). Weekly calibration is manual, using traceable standards (see Section 2.9 earlier). The results obtained are fed into a central computer system. Real-time or near-real-time reports are published either on the World Wide Web (http://www.aeat.co.uk/netcen/airqual/bulletins/) or in the UK on the 'CEEFAX' and 'TELETEXT' systems (bulletin boards transmitted nationally on otherwise unused lines in television broadcasts). Figure 6.18 shows some typical reported readings.

6.3.5 Remote Sensing and other Advanced Techniques

We have seen that many instruments are available for gases which can sample and analyse at locations remote from a laboratory. Spectrometric methods are often used. You may wonder why, if you are simply measuring the light absorption of a gas, you need to take a sample at all. Why not simply measure the light absorption through a section of the atmosphere? This is the principle of *remote sensing*.

Most compounds of environmental concern are found at ppb (v/v) concentrations. Confirm this by looking back at Figure 6.1. You may consider these concentrations **too** low for analysis by direct absorption measurements, but you should remember that extremely long pathlengths can be used to compensate for the low concentrations.

Typical applications are for localized pollution problems, monitoring plumes from chimneys or more general atmospheric surveys within, for instance, an urban area. A more recent use is with roadside monitors to measure the concentration of pollutants in vehicle exhausts. The instruments can be at fixed locations, as

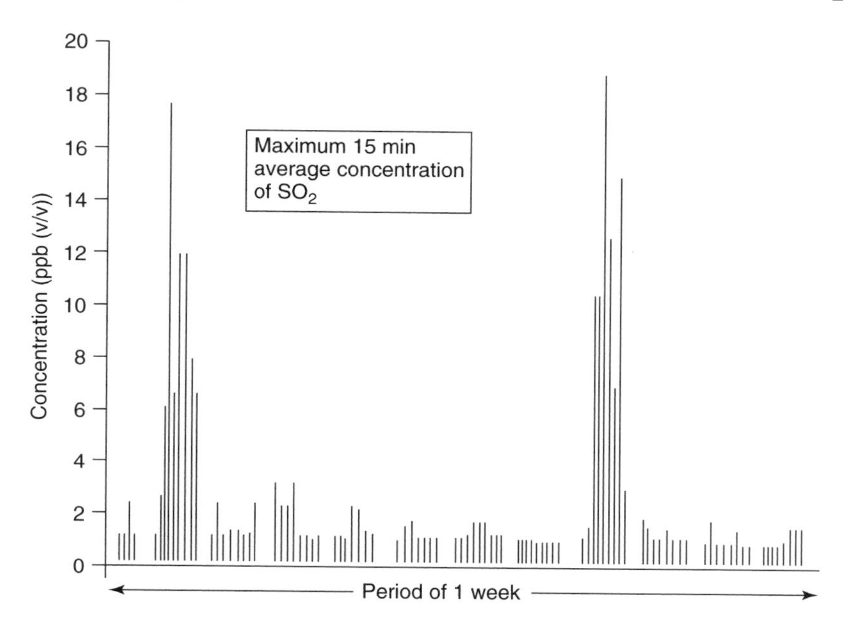

Figure 6.18 Typical Web display obtained from an automatic monitoring network.

part of mobile laboratories or, for the upper atmosphere, from satellites, balloons or aircraft.

The instrumentation needed for this seemingly simple technique is often complex, with much use being made of laser techniques and advanced signal processing.

First of all, we must choose a wavelength of light which is absorbed by the analyte and not by other components in the atmosphere. A number of atmospheric components show characteristic absorptions in the ultraviolet region of the spectrum. These include the following:

Ammonia
Nitrogen dioxide
Radicals such as $OH^{\bullet}$
Unsaturated organic compounds
Mercury
Ozone
Sulfur dioxide

Many of the spectra of such species are highly structured and a small change in wavelength can move from an absorption maximum to a minimum. Figure 6.19 shows a representation of a section of the UV spectrum of sulfur dioxide.

Differential Optical Absorption Spectrometry (DOAS) measures the absorption of UV light over a fixed pathlength, which is typically several hundred metres to several kilometres. The source and detector can be at the same location with a mirror at the end of the sampling path or alternatively the source and detector

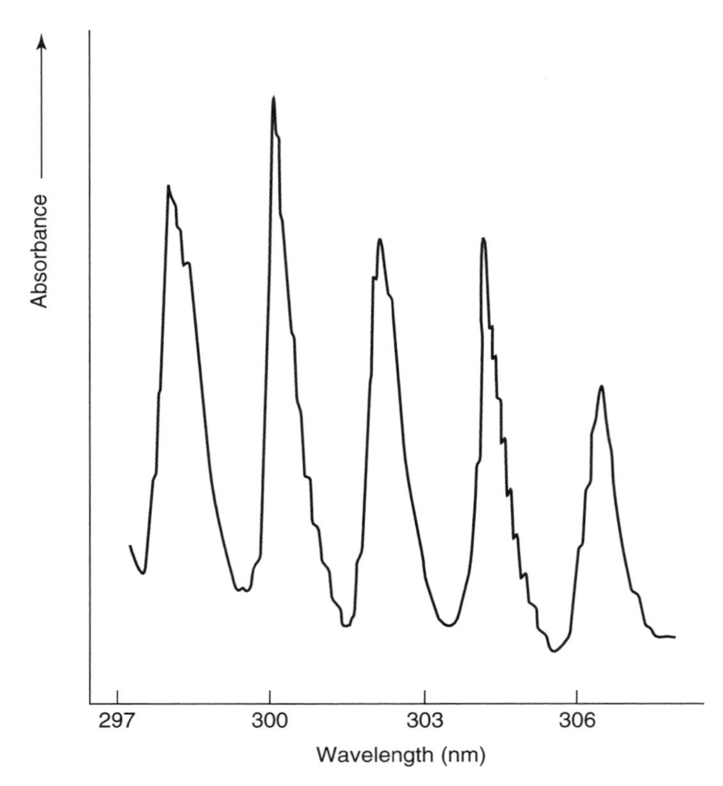

Figure 6.19 Representation of a section of the ultraviolet spectrum of sulfur dioxide.

can be separately located at each end of the sampling path. The first method is currently favoured due to its case of setting up but may suffer from interference by adsorbed contaminants on the mirror surface. The source can be an incandescent or arc lamp, or a laser, with a spectrometer as the detector. The results obtained represent the average value over the whole of the pathlength. Ozone typically has a limit of detection of 4 ppb (v/v) with a 5 km pathlength. Several species can be measured together by monitoring at different wavelengths. A major advantage of this direct measuring technique is that highly reactive species can be measured in the atmosphere, e.g. $NO_3^{\bullet}$, $OH^{\bullet}$, $ClO^{\bullet}$. It is the study of these species which has lead to our current understanding of atmospheric chemistry.

LIDAR (*Light Detection and Ranging*) refers to a family of techniques which can produce a concentration profile within a fixed section of the atmosphere. In its simplest form, the source is a pulsed laser. The light is scattered by particles in the atmosphere, thus providing different light absorption paths before detection. A typical range would be up to a few kilometres. Figure 6.20 shows a configuration where the light detector is positioned close to the laser.

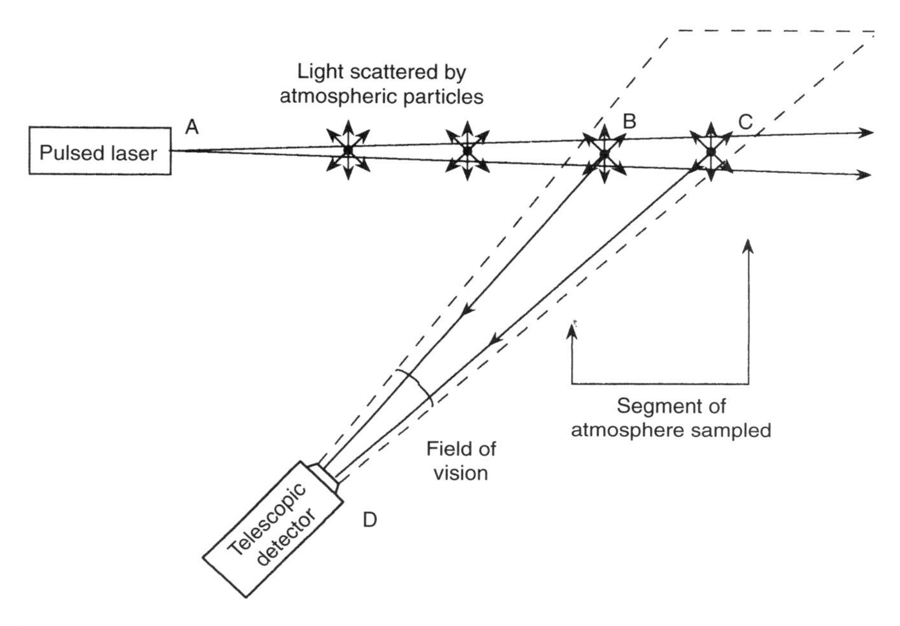

Figure 6.20 A typical configuration of light detection and ranging (LIDAR) used for remote sensing.

The light reaches the detector over a slightly more extended time-period than the original pulse length due to the different pathlengths in the atmosphere. Looking at Figure 6.20, the light travelling on path ABD will reach the detector before the light on path ACD. The intensity of light reaching the detector is measured over the complete return-time period. This information can be processed to give the concentration of the absorbing species over each of the light paths, and this can be built up to give a concentration profile over the complete sampling range.

The most commonly used version of LIDAR is known as DIAL (*Differential Absorption* LIDAR) and uses a pulsed laser at two wavelengths corresponding to the absorption maximum and minimum, i.e. the second wavelength acts as a background absorption measurement.

The infrared region can be used for monitoring local pollution of compounds which do not absorb in the ultraviolet region. More widespread use is hindered by strong absorption of the atmosphere (by carbon dioxide and water) over much of the infrared range, with also the width of the absorption bands at atmospheric pressure leading to much band overlap. A two-laser technique can be used (compare DIAL) at slightly different wavelengths corresponding to the absorption maximum and background. Typical pathlengths would be up to 100 m, with detection limits over this distance in the ppm range.

In order to reduce the problem of band overlap, the infrared absorption needs to be measured under reduced pressure. The narrower bandwidth of the absorption lines makes the technique virtually interference-free and also allows measurements to be made outside of the normal IR window region. The main technique used is known as *Tuneable Diode Laser Absorption Spectroscopy* (TDLAS). In this technique, the atmosphere is continuously sampled by the drawing through of a pressure-reducing valve, with absorption measurements being taken at typically 30 mbar pressure. The technique is often used for single pollutant studies as it is not easy to cover large wavelength ranges with diode lasers. This method as described is obviously not 'remote sensing' in as far as samples have to taken for introduction into the spectrometer. It is, however, frequently used in mobile laboratories alongside the other techniques described in this section.

DQ 6.15

TDLAS has been used in true remote sampling mode for measurements in the upper atmosphere. The spectrometer components and reflection mirror are suspended from a balloon with the absorption path being the atmosphere between the spectrometer and mirror. Why do you think this form of infrared spectrometry can be used for direct atmospheric measurements in the upper atmosphere but not in the lower atmosphere?

Answer

As you proceed up in the earth's atmosphere, the atmospheric pressure decreases. High enough up in the atmosphere, the pressure is low enough to decrease the bandwidth of the absorption lines to remove band overlap, and so potential interferences.

SAQ 6.6

Compare the potential uses of absorption trains, adsorption on to solids and direct-reading instruments as methods of analysis for atmospheric samples.

SAQ 6.7

Which techniques would you use for the following analyses?

(a) Nitrogen dioxide in the external atmosphere at several locations.
(b) An organic solvent in a laboratory atmosphere.
(c) Carbon monoxide, to protect a worker in an area where there may be rapid increases in concentration.

SAQ 6.8

Draw a graph which has as the *x*-axis 'ease of use in field' (scale, easy–hard) and the y-axis 'precision' (scale, low–high), and mark the approximate positions you would place the following:

- Detector tubes
- Passive sampling
- Active sampling
- Electrochemical sensors
- Portable infrared spectrometers
- Portable gas chromatographs
- Remote sensing
- Analytical spectrometers

What conclusions can you draw from this graph?

Summary

Concern over gaseous pollutants includes not only those found in external (outdoor) atmospheres, but also internal (indoor) atmospheres, since both can have an effect on human health. The types of pollutant found in these two areas can differ in chemical type and in concentration, with higher concentrations often being found with internal atmospheres. The concentrations can often change rapidly with time. Concentrations averaged over a fixed time-period (time-weighted averages) are the most appropriate measurements for long-term investigations. A detailed study of a pollution incident would, however, require instantaneous concentration measurements. Methods have been described for both of these types of measurement, and include techniques which determine concentrations directly in the field (including remote sensing), and techniques requiring the analysis to be completed in the laboratory. Use of such methods for flue gas analysis is also described.

Chapter 7

Atmospheric Analysis – Particulates

Learning Objectives

- To understand the importance of particulate material in the atmosphere and the importance of particle size.
- To determine appropriate methods of sampling particulates in external and internal atmospheres and in flue gases.
- To assess the relative merits of analyses involving sample dissolution and methods not requiring a dissolution stage.
- To appreciate the range of solid-state analytical techniques which may be available in specialist laboratories.

7.1 Introduction

Have we now discussed analysis of all of the major components of the atmosphere? Certainly not! We have only looked at gases and vapours so far. An equally important component is the particulate matter. Particulates are a natural component of the atmosphere. They include the following:

- Condensation products from natural combustion (forest fires, volcanoes, etc.).
- Products of reaction of trace gases (ammonium chloride, sulfate and nitrate salts).
- Material dispersed from the earth's surface (salt spray from oceans and mineral dust from continental land mass).

In addition to these is the particulate material introduced by man. This can predominate in urban atmospheres, with the major sources being combustion and incineration processes.

Particulates have an important role in the chemistry of the atmosphere. Many atmospheric reactions which, at first sight, appear to take place in the gas phase, occur either on the surface of the particulate matter or in the liquid phase – in water adsorbed on to the surface of the particle. Let us take as an example the smogs which regularly affected London until the mid-1950s. The primary components of the smog were sulfur dioxide and particulate matter, both derived from coal combustion. In one major incident in 1952, approximately 3000 deaths in one week were attributable to the smog. The maximum concentration of sulfur dioxide was found to be 3.8 mg m^{-3} (1.34 ppm). This concentration, which was very much larger than normal, has been shown to cause no adverse effects in man without the presence of particulate matter. A strong synergistic effect was indicated. The particulate matter provided a surface for the liquid phase oxidation of sulfur dioxide to sulfuric acid, which remained adsorbed on the surface of the particle. The size distribution of the particulate matter was such that, on inhalation, a fraction of the particles lodged in the lungs. The irritation caused by the particulate material was increased by the adsorbed layer of sulfuric acid.

Atmospheric transport in the form of particulates is one of the major methods for the dispersal of pollutants. We have seen earlier in Chapter 2 that a significant route for dispersal of lead is via the atmosphere (the lead being transported predominantly as inorganic salts). Similar routes can be constructed for the other metals of environmental concern. Semi-volatile organic material occurs in the atmosphere partly in the vapour state and partly in the solid phase, either as organic particulates or adsorbed on to inorganic particulates. This category would include most pesticides. Any consideration of the transport of these organics needs to take both vaporized and particulate fractions into account.

Let us now consider which measurements may be useful for characterization of the particulate content of an atmospheric sample:

(i) A preliminary measurement would be the total particulate concentration. This is a measurement of the weight of solid extracted from a fixed volume of the atmosphere by filtration or by other methods (see Section 7.2 below). Typical values are as follows:

   70 $\mu g\ m^{-3}$ rural air
   300 $\mu g\ m^{-3}$ urban air
   10 mg m^{-3} factory workshop air
   100 mg m^{-3} power station flue gases

(ii) The second consideration is the analytical composition. For metals, this is often simply elemental analysis. The analytical task can be more difficult than we found for aquatic samples since the inorganic component of the particulate material may be highly insoluble, particularly if present as silicate salts. All of the analytical techniques we have so far come across involve sample dissolution. Two approaches are possible for 'insoluble' particulates. Extreme conditions may be used to dissolve the sample, followed by the analytical methods for metals which we have already discussed. The alternative approach is the use of techniques which do not require sample dissolution. Organic analysis is generally simpler, involving dissolution, followed by the analytical techniques already discussed.

(iii) The particle size distribution is often also determined.

DQ 7.1

Why do you think particle size is important?

Answer

1. ***Transport***. *The residence time of a particle in the atmosphere is dependent on its size. The greater the size, then the more rapidly deposition from the atmosphere occurs (see Table 7.1 below). Particles less than 0.1 μm diameter can be considered to be capable of permanent suspension.*
2. ***Differences in Physiological Properties***. *The smaller the particle size, then the greater is the possibility of the particle entering the gas-exchange region of the lungs. It is this material which will have the greatest potential physiological effect. This fraction of the particulate matter is termed 'respirable' dust, and as a guide would refer to material below approximately 5 μm. The larger fraction which enters the nose and mouth during breathing is known as 'total inhalable dust'.*
3. ***Distribution of Chemical Species***. *If you are studying emissions from a particular industrial process, you may find that the particulate matter is often within a narrow size range. Fractionation of the dust sample may then constitute an essential part of the analytical procedure.*

Table 7.1 Classification of particulate material

Particulate diameter (μm)	Commonly used term	Sedimentation velocity[a] (cm sec^{-1})
<0.1	Fume	Negligible
0.1–10	Smoke	8×10^{-5}–3×10^{-1}
10–100	Dust	0.3–25
>100	Grit	—

[a]In still air.

Particle size measurements will be important in determining the most suitable method of pollution control.

4. ***Effect on Atmospheric Reactions.*** *We have seen that many reactions take place on the surface of particles. The surface area per unit mass decreases with an increase in particle size for similarly shaped particles.*

Of major current concern for external atmospheres are the particles with an aerodynamic diameter less than 10 μm, known, even in the popular press, as PM_{10}s. These can occur from a number of sources. Road transport can contribute up to 50% in an everyday city atmosphere but can be a much larger percentage when the concentrations exceed statuary levels. Other sources include power stations and coal fires. The levels in built-up areas correlate quite strongly with the numbers of respiratory disorders. The PM_{10} concentrations at the UK National Air Quality Objective of 50 $\mu g\ m^{-3}$ (running 24 h mean) have been estimated at producing one additional hospital admission per day (for respiratory disorders) per million of population. There appears to be no safe concentration below which health effects are absent.

Due consideration has to be taken in all of the following methods of the low particulate concentrations found in the atmosphere. Even with long sampling times in heavily polluted atmospheres, you will only be dealing with milligram quantities of sample, thus making a very exacting analytical task. As in other sections, we will be discussing sampling methods first, and then the analytical methods. The analytical methods are divided into those requiring sample dissolution prior to analysis and those which can analyse solid material directly without a dissolution stage. The solid-phase analytical techniques are then briefly described.

7.2 Sampling Methods

The importance of careful sampling strategies has been stressed throughout this book, but perhaps nowhere is it more important than with particulate sampling. Concentrations can vary rapidly with time and location. In internal atmospheres, there is often measurable vertical variation, even over a few centimetres. This leads to an emphasis towards personal sampling to assess the exposure of an individual rather than comprehensive surveys of background levels. For external atmospheres, however, measurement of background concentrations using large-throughput (high-volume) samplers, remains the most appropriate method.

7.2.1 High-Volume Samplers

In this method, the air sample is drawn through a large-diameter membrane filter (20–25 cm), typically at 75 $m^3\ h^{-1}$. The construction of the sampler (Figure 7.1)

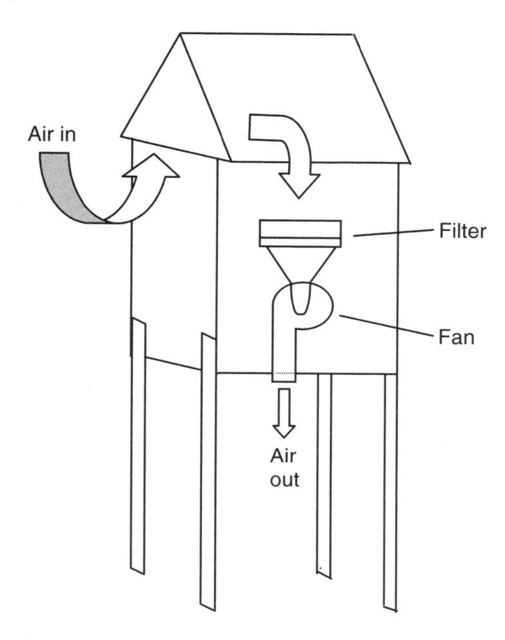

Figure 7.1 Schematic of a high-volume sampler with shelter.

is most easy to understand when you discover that the earliest types were modified from commercial vacuum cleaners, i.e. it is simply a fan behind a filter holder. Nowadays, of course, purpose-built apparatus is readily available. Typical sampling times range from 1 h for contaminated urban atmospheres to 12 h for clean rural atmospheres, with shorter times possible for internal atmospheres.

The choice of filter is based on the following:

- Retention of correct particle size range.
- Absence of trace impurities in the filter.
- Compatibility with the subsequent analytical procedure. Some procedures require the total combustion of the filter, and others its dissolution.

Cellulose filters should be used for metals and inorganic anions, and glass-fibre filters (or under some circumstances, silica filters) for organics.

7.2.2 Personal Samplers

With this sampler, a filter-holder is clipped to the lapel, with the pump around the waist. The pump is similar in design to those used for organic gas sampling as described in the previous chapter, with one difference – dust sampling is at the higher rate of approximately 2 l min^{-1} through a 25 mm filter. Filters are made of glass fibre if simply a total particulate weight is required. Other filter material may be used for elemental analyses, the choice depending on the subsequent analytical procedure.

This equipment will produce a representative sample of total inhalable dust. If a sample of respirable dust is required, then a pre-selector is necessary to ensure that only particulates of the correct size range reach the filter. A cyclone elutriator (Figure 7.2) may be used for this purpose. In the latter, the gas is spiralled through a conical container in such a way that particulates outside the required size range fall into a container at the base of the elutriator, rather than passing on to the filter.

7.2.3 Cascade Impactors

The previous two methods have used filtration for collection of the particulate material. Cascade impactors rely on adhesion of particulates on to a surface. The particulates are fractionated according to their mass. A typical apparatus is shown in Figure 7.3. In this, air is drawn through the device at a constant rate to impact

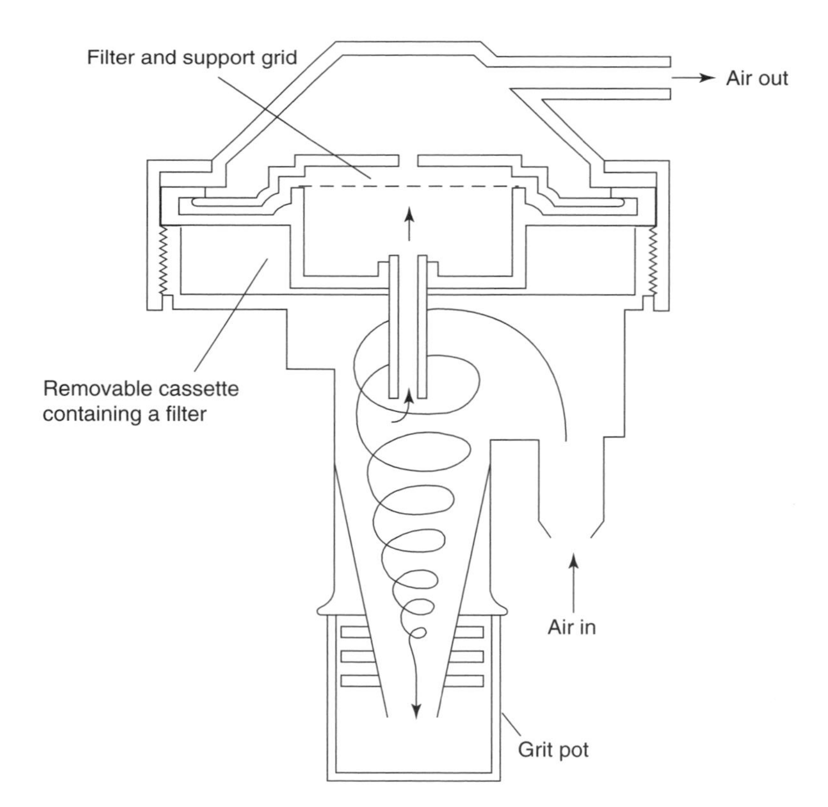

Figure 7.2 Schematic of a cyclone elutriator.

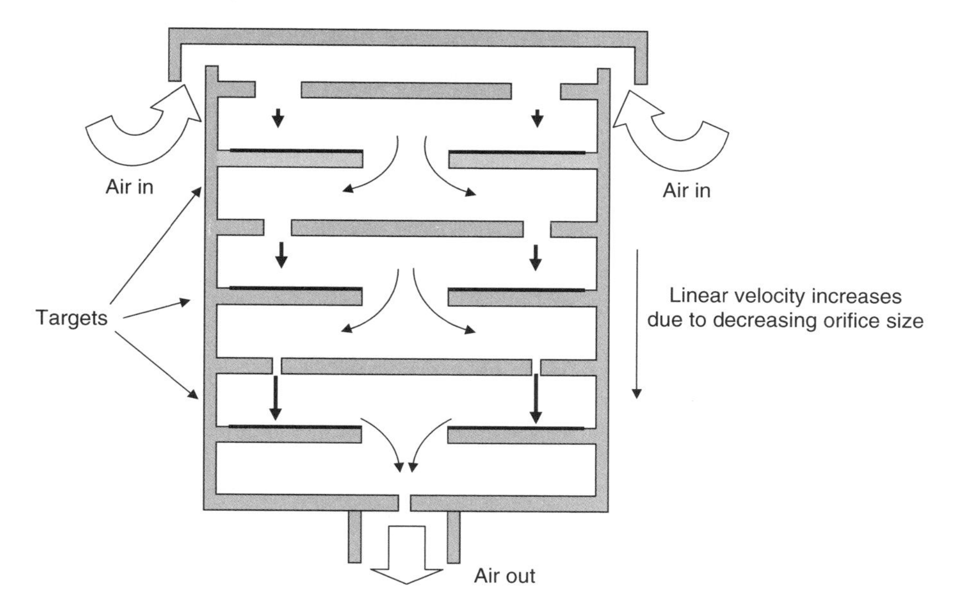

Figure 7.3 Schematic showing the operation of a cascade impactor.

on a number of targets coated with petroleum or glycerine jelly. By constriction of the flow before each target, the linear velocity of the air increases. Particles adhere to the targets if they impact above a specific momentum (momentum = mass × velocity). Since the air velocity increases through the apparatus, successively smaller particles will adhere to each successive surface. A typical operating range is 0.5–200 μm.

Typical operating flow rates would be 1 m^3 h^{-1}, producing only a few micrograms of sample in each fraction per hour of operation when sampling a typical urban atmosphere.

7.2.4 Further Considerations for Organic Compounds

If sampling is to represent the total organic content of the atmosphere, it has to accommodate both solid and vapour phases. After passing through the sampler to determine the particulate matter, the gas is drawn through an adsorbent to extract the vapour-phase component. A polyurethane foam cartridge is often used as these produce a lower pressure drop than the adsorbents discussed earlier in Section 6.2. This permits the higher sampling rate necessary for the particulate analysis. One manufacturer produces a filter and two-section adsorbent in a single unit. The analysis of the two phases then proceeds separately with extraction of organics from the filter typically being by using a Soxhlet apparatus (see Figure 5.1 earlier).

7.2.5 Sampling Particulates in Flowing Gas Streams

This is normally carried out by using a filter included in a sampling train designed either specifically for particulates or as a combined train for particulates and gases (see Section 6.2.1, and below).

7.2.5.1 Isokinetic Sampling

There is an important consideration when analysing particulates above about 5 μm in diameter. By sampling a gas, you are in fact disturbing the flow patterns of the gas itself and this may lead to errors in the measured particulate concentration. If the sampling rate is quicker than the gas flow, then the flow pattern will be distorted and will bend into the sampler. The particulate material will have a greater inertia than the gas molecules. It will tend still to travel in the original direction and so will not enter the sampler. The measured particulate concentration will be less than the correct value. If, however, the flow is less than that of the sampled gas, then the gaseous molecules will be diverted around the sampler. The particulate content will tend to travel directly into the sampler. In this case, the analytical value will be greater than the true value. The least disturbance is when the sampling flow rate is identical to the gas flow rate (Figure 7.4). This is known as *isokinetic sampling*.

The sampling apparatus in a flowing gas stream will normally also include a *pitot tube* which measures the linear velocity of the gas stream. This will enable the flow rate of the sampling to be matched to that of the gas. It is also used in preliminary investigations to determine the flow patterns in the gas stream. One design of pitot tube is shown in Figure 7.5. The pressure difference is measured by a manometer between two ends of the tube, with one pointing directly into the flow and the other in the opposite direction (i.e. along the flow). The pressure difference is proportional to the square root of the linear velocity.

The flow problems with particulate sampling have implications in the construction of other samplers discussed elsewhere in this chapter. A poor internal construction within a sampler may inadvertently discriminate against particular particle sizes. From a more beneficial point of view, selection of particular sizes can be achieved by suitable inlet design.

7.2.5.2 Design of Sampling Train

Flue and exhaust gases are invariably at a higher temperature than ambient. If the particulate material is not sampled at the flue temperature, condensation of the water or other vapour-phase components could occur. This would lead to blockage of the sampling train, as well as inaccuracies in the analytical measurement. There are two approaches to overcome this problem. One approach, which is used in at least one European standard method, is to have the filter inside the flue itself. The size of the filter apparatus can, however, affect the flow patterns within the flue. An alternative approach, used in EPA Method 5, has the filter in

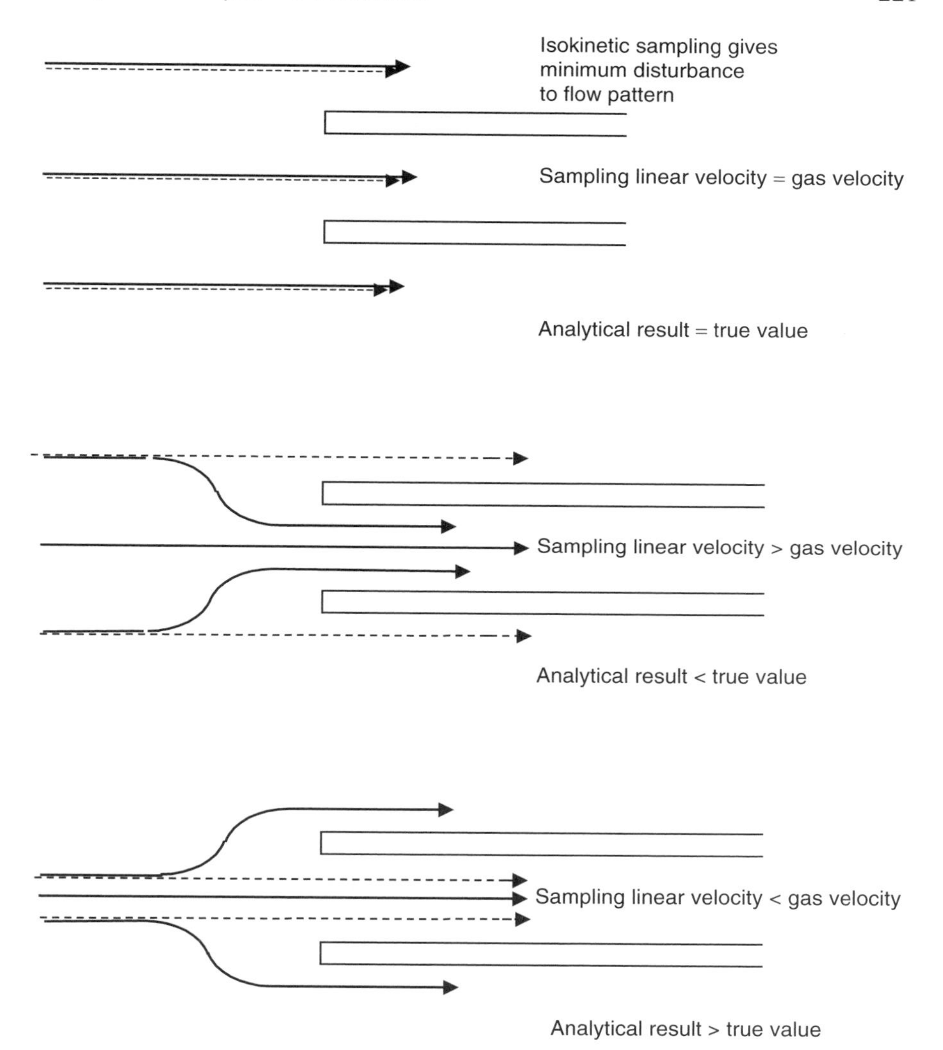

Figure 7.4 Ilustration of isokinetic sampling: (⟶), gasflow; (--->), particulate flow.

a heated compartment outside the flue and samples the flue gas by using a heated sampling probe. This gives lesser disturbance of the gas flow, although there may still be condensation problems if the temperatures are not correctly matched.

Sampling ideally needs to be in several locations in the flue. The filters are usually of quartz or glass fibre and, for low particulate concentrations, standard filter discs may be used. The filter may be in the form of thimbles for higher concentrations. These give the highest surface area for collection of the

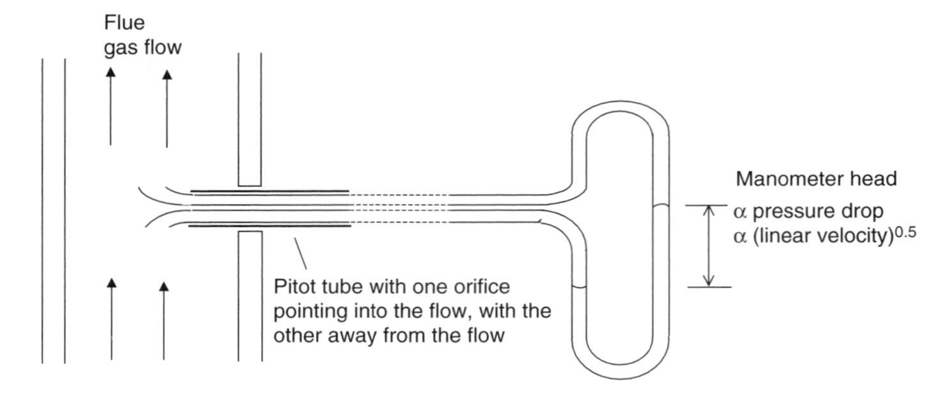

Figure 7.5 Ilustration of a pitot tube in use.

particulates and also minimize particle loss in handling. They are often placed after a cyclone separator (see Section 7.2.2 above) which removes the larger particles.

7.2.6 PM_{10} Sampling

Concern for PM_{10} measurements has predominantly been for external atmospheres but personal monitoring is also important to determine individual exposure. A number of instrument designs are available. Many are based on filters, with the inclusion of a pre-selector which only allows the PM_{10} fraction to be collected. The PM_{10} pre-selector may use cyclonic or impaction techniques (see Sections 7.2.2 and 7.2.3 earlier). Optical methods are also available for PM_{10} measurements which measure the atmospheric light scattering from the suspended particles. Filter methods (e.g. the Partisol Air Sampler) are used as the reference method for batch sampling. The Partisol instrument automatically changes the filter every 24 h. The filters are then weighed after equilibration with the atmosphere at room temperature. Other instruments can be used for continuous monitoring. Two examples are described below.

A *Tapered Element Oscillating Microbalance* (TEOM) instrument is shown in Figure 7.6. The atmosphere is passed through a heated filter (50°C) at the end of a tapered oscillating glass tube. The change in oscillation frequency as the gas passes through is directly related to the mass of particulate material accumulated on the filter. This method has been chosen for the UK urban monitoring programme as it is one of the few methods being able to provide hourly readings necessary for the UK standard which is based on a 24 h rolling average.

The *β-attenuation* instruments collect the particulate matter on a moving strip of filter paper after the PM_{10} selector. Here, β-rays from a radioactive

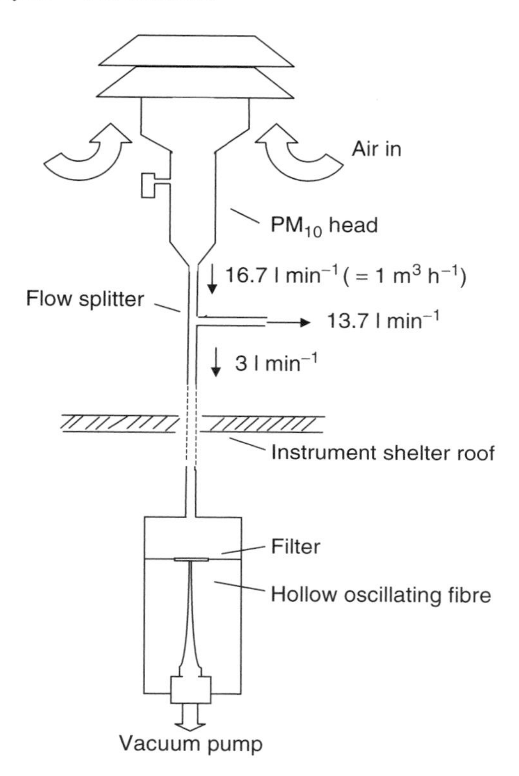

Figure 7.6 Schematic of a tapered element oscillating microbalance with PM_{10} head.

source are passed through the filter, with the absorption of the radiation being proportional to the mass of particulate matter on the filter. The inclusion of a radioactive source could mean that for some applications the microbalance may be preferred.

DQ 7.2

There are currently several investigations proceeding to compare the results of different PM_{10} analysers. What reasons could contribute to this concern?

Answer

Any atmospheric particle measurement will be dependent on atmospheric conditions. The critical conditions could include wind direction and velocity with respect to the sampling head, humidity and ambient temperature (which would affect the collection and retention of volatile components). The influence of these factors will be instrument-design-dependent.

The instrument designs involve a wide range of physical techniques from filtration to optical scattering. Each technique will respond differently to individual particle sizes, thus potentially leading to a slightly different response with each instrument. You should remember that particulates in the atmosphere always occur over a wide range of sizes.

Even with instruments based on the same principle, the detailed internal construction may give inadvertent selectivity due to kinetic sampling problems (see Section 7.2.5 above).

The TEOM instrument collects samples on heated filters. Some components, such as ammonium compounds (see Section 7.1) and particulate VOCs are volatile. There is therefore the potential that these components may be lost at the elevated temperatures.

7.2.7 Sampling of Acid Deposition

The chemistry of acid rain is complex, involving not only gaseous acidic compounds but also particulates. Within cities, the major component is particulate material and this is known as 'dry deposition'. Acid rain whose effects can be traced over long distances (hundreds of miles) is predominantly gaseous and this is known as 'wet deposition'.

Any instrument designed to monitor acid rain will have to be able to collect both types of deposition and determine them separately. Figure 7.7 shows an instrument containing two sample containers which are automatically opened or closed according to the rainfall. The aqueous samples collected can be analysed by the standard methods already discussed. Typical analyses would include common anions (measured by UV/visible spectrometry or ion chromatography) and common metal ions (measured by flame photometry or ion chromatography). The sample collected on the dry side must be dissolved or suspended in distilled water before analysis.

SAQ 7.1

Consider a studio glass-making furnace in a small room for which there is concern over a technician's exposure to the particulate lead emissions. How would you set about sampling the atmosphere?

SAQ 7.2

List components which may be needed to determine both particulate material and gaseous components in a flue gas by using a combined sampling train.

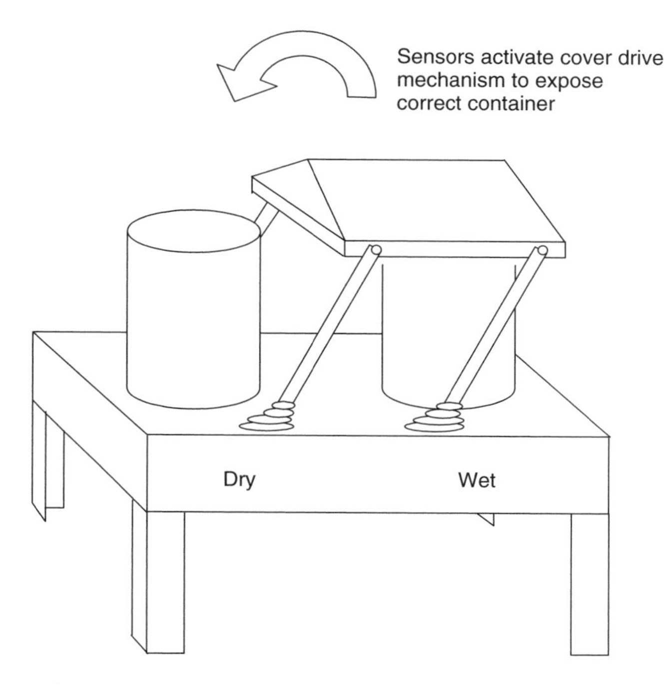

Figure 7.7 Schematic of a sampler used for wet and dry acid deposition.

7.3 Analytical Methods Involving Sample Dissolution

7.3.1 Metals

A first step in any analytical procedure should be to consider the probable composition of the sample. This is a vital step for particulate analysis which will allow the correct choice of dissolution technique. If the composition of the sample is unknown, as would be the case for many external atmosphere samples, hydrofluoric acid, which is capable of dissolving silicates, may be required. This acid causes severe burns and attacks glass apparatus (the silica structure of the glass is closely related to the insoluble silicates which you may be trying to dissolve). 'Teflon' apparatus is required and the analyses should be performed in a hydrogen fluoride-resistant fume cupboard. You may now be able to see why this method is avoided whenever possible.

If the composition of the dust sample is known (as may be the case with samples from workplace environments), the dissolution may be less severe, according to the known solubility of the sample. Dilute acid, mild oxidizing agents, or even water may be all that is necessary for dissolution.

To illustrate the difference, let us look at two standard methods for the analysis of lead in dust:

1. The SCOPE procedure of the International Council of Scientific Unions which involves the following stages:

   - Collection of the particulate matter in a glass-fibre filter (twice washed with distilled water).
   - Warming with hydrofluoric acid until the liquid is almost evaporated.
   - Repeating with nitric acid.
   - Making up to volume with distilled water.

This procedure solubilizes the filter as well as the sample.

2. The UK Methods for the Determination of Hazardous Substances procedure (MDHS 6/3) used for internal atmospheres assumes that the lead is in a more easily soluble form and employs a simpler one-stage procedure of warming with nitric acid–hydrogen peroxide. The filter paper remains undissolved. Once the sample is dissolved, the analysis can proceed by a number of methods available for determining metal ions in solution.

DQ 7.3

Which two methods have you come across which may be most suitable for routine analysis of metals in particulates?

Answer

(i) Atomic absorption spectrometry
(ii) Ultraviolet/visible absorption spectrometry

For less routine analysis, and particularly for analysis of metals at low concentrations, other techniques may sometimes be used. These include inductively coupled plasma-optical emission, inductively coupled plasma-mass spectrometry, flame atomic emission and atomic fluorescence techniques. Ion chromatography can also be used for common main group metal ions (Na^+, K^+, Ca^{2+} and Mg^{2+}) as well as for common anions.

The sensitivity of each technique is different for each element. Some comparative data are shown in Table 7.2. Take care, though, when using such a table as limits may change between equipment manufacturers and with improvements in instrumentation. If you assume a 1 m^3 air sample with the metal extracted into 5 ml of acid, the table covers a range of detection limits from 4 $\mu g\ m^{-3}$ (Cd using atomic emission) to 5×10^{-6} $\mu g\ m^{-3}$ (Ca using atomic fluorescence and Pb using ICP-MS).

7.3.2 Organic Compounds

Simple determination of organic content may be by analysis of total organic carbon (see Section 3.3.3 above) or by weight loss after extraction with an

Table 7.2 Comparative detection limits ($\mu g\ l^{-1}$) of atomic spectrometric techniques

Element	Furnace AAS	ICP-OES (axial viewing)	ICP-MS	Flame atomic emission	Atomic fluorescence
Ca	0.05	2	2	0.1	0.001
Cd	0.003	0.2	0.003	800	0.01
Mn	0.01	0.1	0.002	5	2
Pb	0.05	0.8	0.001	100	10

organic solvent. The components of the extract can then be determined by the chromatographic and spectrometric methods described earlier in Chapter 4.

SAQ 7.3

Atomic absorption and ultraviolet/visible spectrometry are often specified as alternatives in standard methods for the analysis of metals as particulates in workplace atmospheres. This contrasts with the predominance of atomic spectrometric techniques for the analysis of aqueous samples. What are the possible reasons for this difference?

SAQ 7.4

Which feature of the sampling and analysis of atmospheric particulates could account for the large number of different techniques used for the analysis of low concentrations of metals?

7.4 Direct Analysis of Solids

We will now briefly discuss a number of representative techniques for direct analysis. The first three are methods for elemental analysis using equipment which will only be available in specialist laboratories. An example of a solid-state method using a readily available laboratory instrument is then discussed. The final section briefly mentions the specialized techniques used for asbestos analysis.

7.4.1 X-Ray Fluorescence

This technique is based on the irradiation of an atom with X-rays leading to the ejection of an electron from an inner shell. Outer-shell electrons cascade to the inner shell to fill the vacancy, emitting X-rays. The wavelength of this radiation is related to the atomic number of the nucleus according to the following equation:

$$\frac{1}{\lambda} = kZ \tag{7.1}$$

where λ is the wavelength of radiation, k is a constant and Z is the atomic number.

Elements thus emit radiation at characteristic wavelengths. Absorption and emission occurs predominantly in the first few surface layers of atoms. With suitable corrections for matrix effects, which may include the preparation of standards with compositions as close as possible to the sample, the intensity is proportional to the concentration of the element. Samples can be in either the liquid or solid state; hence the use of the technique for contaminated land and waste analysis, as discussed above in Sections 5.5 and 5.6.

Two types of instrument are available, which differ according to how the fluorescent radiation is analysed. Wavelength-dispersive instruments measure the radiation at each wavelength sequentially, using diffraction from a rotating crystal to direct specific wavelengths to the detector (Figure 7.8).

Energy-dispersive instruments measure the whole of the fluorescence simultaneously at the detector. The contributions from each wavelength are then separated electronically. This type of instrument is more convenient to use and produces more rapid analyses, but has a slightly lower sensitivity.

A typical X-ray fluorescence spectrum is shown in Figure 7.9. Elements above atomic number 40 can be routinely analysed. By using vacuum techniques, which prevents absorption of X-rays by low-atomic-mass elements in the atmosphere, elements from F to Ca can also be measured. Particulate samples collected on filter paper can be analysed without any pretreatment being required. For routine analysis, concentrations are determined using calibration filters, with quality control procedures including occasional cross-checking with an atomic absorption or ICP analysis. Detection limits for elements vary widely, but for airborne particles they are of the order of 10^{-2} $\mu g\ m^{-3}$, when expressed as the original atmospheric concentration.

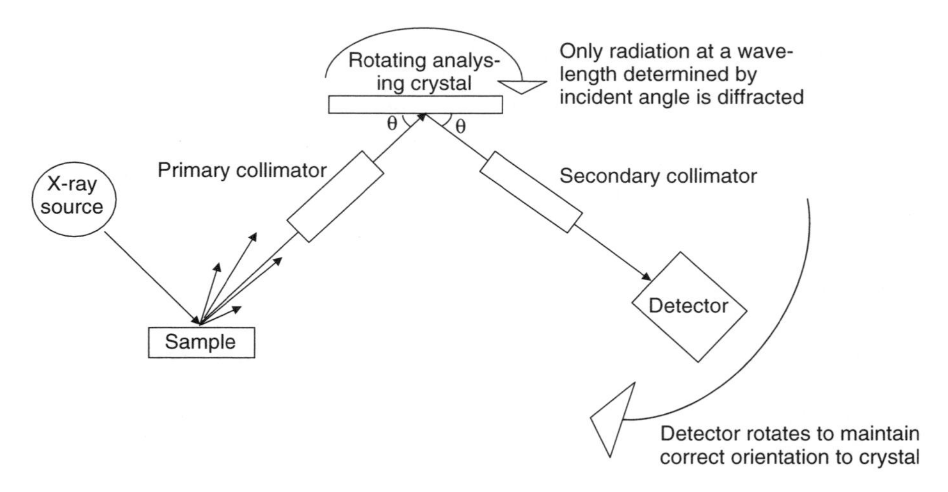

Figure 7.8 Schematic of the components of a wavelength-dispersive X-ray fluorescence spectrometer.

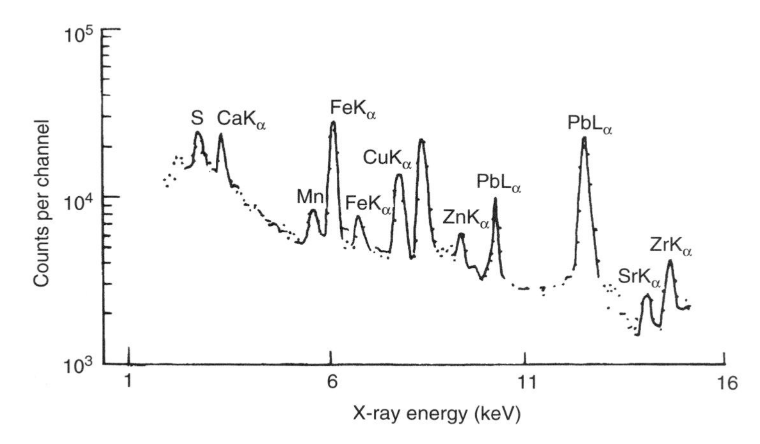

Figure 7.9 X-ray fluorescence spectrum of a dust sample. From Manahan, S. E., *Environmental Chemistry*, 6th Edn. Copyright Lewis Publishers, an imprint of CRC Press, Boca Rator, Florida.

Instruments are also available for use outside the laboratory. These are energy-dispersive spectrometers. They can range from transportable instruments for mobile laboratories to hand-held instruments for specific elements. In the smaller instruments, the X-rays are produced from a small radioactive source, e.g. ^{109}Cd. Larger transportable instruments (and also laboratory spectrometers) generate X-rays by the acceleration of electrons on to a chromium or tungsten target.

The hand-held instruments can be used for rapid assessment of elements on particulate filters. As it is assumed that the absorption takes place on the surface layer, there is no need for matrix correction. The same instrument can also be used for contaminated site monitoring (see Section 5.5 earlier), where the instrument is simply placed on the flat surface of the suspect ground, and also waste monitoring (see Section 5.6 above).

7.4.2 X-Ray Emission

X-rays may also be generated by the bombardment of a sample with fast electrons. The bombardment again causes excitation of inner-shell electrons with the subsequent decay back to the ground state causing the X-ray emission. This technique is used in the electron microprobe analyser. The electron beam can be focused on a small area, which can be as small as an individual dust particle. This then gives an extremely powerful technique in assessing a composite dust sample. It is also one of the few techniques capable of quantitative analysis of the low-mass samples produced by a cascade impactor. The instrument may also be used as a conventional microscope. Qualitative analyses of the particles can be made from the images produced, emitted by X-rays at wavelengths corresponding to individual elements.

7.4.3 Neutron Activation Analysis

In this technique the sample is irradiated with neutrons to produce radionuclides of the elements of interest. As the radioactive nucleus decays, it emits gamma rays. The intensity of the gamma ray spectrum can be related to the original concentration in the sample, for example:

$$^{55}_{25}\text{Mn} + ^{1}_{0}\text{n} \longrightarrow ^{56}_{25}\text{Mn} \longrightarrow ^{56}_{26}\text{Fe} + \beta^{-} + \gamma(t_{1/2} = 2.58\ \text{h}) \qquad (7.2)$$

The technique is highly sensitive, needing as little as 0.1 μg of sample. Detection limits for elements in airborne particulates can be as low as 2×10^{-5} μg m^{-3}. No chemical pretreatment is necessary and the only physical treatment required is grinding and homogenizing large samples. The one major disadvantage is that you need a source of neutrons, usually a nuclear reactor!

7.4.4 Infrared Spectrometry

This method can be used for compounds which have characteristic absorption frequencies well away from those of likely interfering components. Quartz may be determined by this method using absorptions at 780 and 800 cm^{-1}. The sample is introduced into the beam either directly on the filter paper or after making a pressed disc by grinding the sample with potassium bromide and compressing under high pressure. The absorptions are compared with standards produced from atmospheres containing known quantities of quartz of similar particle size to that of the sample.

7.4.5 Methods for Asbestos Analysis

Asbestos is a term used for any one of a group of fibrous silicate minerals. These materials possess good heat and electrical insulation properties and have found widespread use in industry. It has, however, become a major environmental hazard. Airborne fibres are capable of being trapped in the lungs. The respiratory disease, known as asbestosis, can result, as well as a number of forms of cancer.

The detrimental effects on human health are related to the shape and size of individual fibres. Microscopic analysis is essential. The method using optical microscopy involves collection of the particulate material from the atmosphere by filtration, preparation of a microscope slide and then identifying and counting fibres in the microscope field of view. The latter is then altered and the counting repeated a number of times. The results are expressed in terms of the number of fibres per millilitre of air.

Electron microscopic techniques may also be used, giving the possibility of chemical analysis of individual fibres (see Section 7.4.2 above) as an additional means of identification.

DQ 7.4

Solid-state analytical techniques appear to offer many advantages in the laboratory over methods requiring sample dissolution. What are the disadvantages which sometimes restrict their use?

Answer

1. *The techniques avoid the dissolution stage of other procedures. For particulate analysis, this can be a difficult process. Sample preparation is, however, still required for some solid-state techniques and also in the production of calibration samples. For small numbers of samples, the time saving is then not as great as would first appear.*
2. *The direct analysis of solid material also poses problems for the analysis of large samples. The material analysed (a few milligrams at most) has to be representative of the whole. We have seen, however, that this can be put to advantage with electron microprobes which are able to analyse individual particles.*
3. *Some methods, including X-ray fluorescence, only respond to the first few layers of atoms within a sample. Surface layers may have a different composition to the bulk and, without due care, misleading results may be produced. Another problem with the X-ray method is the possibility of matrix effects.*
4. *Many of the techniques require highly specialized spectrometers which may not be routinely found in general analytical laboratories.*

SAQ 7.5

What criteria would you use to choose the analytical technique for several metal ions in a particulate sample?

Summary

Particulate material is an essential and natural component of the atmosphere. Much airborne pollution is, however, also in the form of particulate material. The particulate size is an important consideration as well as its chemical analysis. An analysis of the material starts with sampling from the atmosphere. This is often by filtration. The method used for the chemical analysis depends on the ease of solubility of the material. If the substance is readily dissolved, then the analysis can proceed using techniques already discussed for species in solution. If the substance is more difficult to dissolve, then techniques which do not require sample dissolution (solid-state analytical techniques) may be used. Examples of both types of method have been discussed.

Chapter 8
Ultra-Trace Analysis

Learning Objectives

- To understand the problems presented by analyses of organic compounds (particularly polychlorinated dibenzo-*p*-dioxins and related compounds) at concentrations of ng kg^{-1} or below.
- To extend the pretreatment and clean-up methods learnt for mg l^{-1} concentrations to these lower concentrations.
- To be able to describe the use of GC–MS for analysis of these compounds, including the application of isotopes and isotopic standards to quantification and quality control.

8.1 Introduction

So far, our discussions have concerned methods using instruments which are commonly available in analytical laboratories. Although you may not believe it when you first try to handle concentrations at μg l^{-1} or μg kg^{-1} levels, these methods can be readily performed by skilled analysts. This final chapter lowers the concentration range studied by a factor of 10^3 or more. We will start the discussion of the analytical techniques from the point of view of 'How do we modify existing methods to gain the required sensitivity?'. We soon find ourselves dealing with instruments which may not be as readily available as those described in the previous chapters. The analyst will need to be highly experienced in order to understand the problems when working at such low concentrations. We have now reached the level at which only a few laboratories in any one country have the necessary skill, expertise and facilities to perform the analysis with accuracy and precision.

DQ 8.1

Why do we need to measure such low concentrations?

Answer

1. *Concentrations can be greatly increased in an organism compared to the environment in which it is living.*
2. *Many of the compounds of concern are suspected to have high chronic and/or acute toxicity.*
3. *Some, but by no means not all, of the compounds of concern are thought to be completely man-made and so any detectable concentration gives an indication of environmental contamination.*

I hope you didn't have any problems with the answer. If you did, you should revise Chapters 1 and 2 before proceeding any further.

8.1.1 What Groups of Compounds are We Discussing?

At the lowest levels of detection, one area of concern is centred around polychlorinated dibenzo-*p*-dioxins (PCDDs). The most well-known member of the group is 2,3,7,8-tetrachlorodibenzo-*p*-dioxin, which shows very high acute toxicity for some species in laboratory tests. Also of concern is the related polychlorinated dibenzofuran group of compounds (PCDFs), the most toxic member once again being the 2,3,7,8-tetrachlorinated compound. These compounds are usually found in the environment in complex mixtures containing PCDDs and PCDFs with all possible substitution patterns (Figure 8.1). I won't ask you to draw all 210 compounds, but you can check that there are 17 PCDDs and PCDFs which include the 2,3,7,8-substitution pattern.

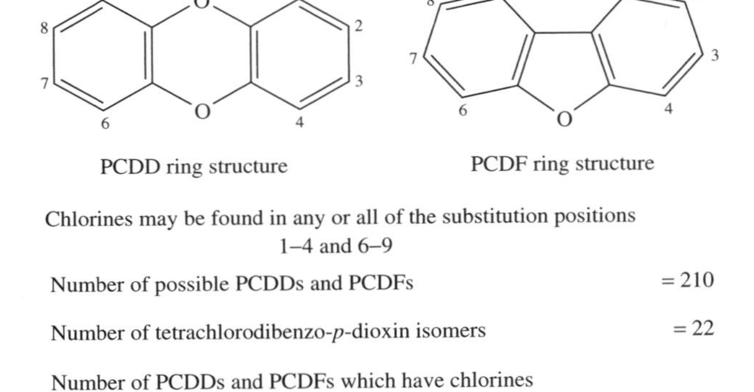

Figure 8.1 Structures and substitution patterns of polychlorinated dibenzo-*p*-dioxins (PCDDs) and polychlorinated dibenzofurans (PCDFs).

The major input of these compounds is from the combustion of organic material containing chlorine. They may also be found as contaminants in some chlorinated chemical products. Combustion sources include chemical and municipal incinerators, coal-fired power stations and domestic coal fires. Part of the ongoing debate over hazardous waste disposal (landfill, incineration or other means) is centred around the toxic products which may be produced by incineration. The compounds also appear capable of being produced naturally by forest and moorland fires.

Polychlorinated biphenyls (an example was shown earlier in Figure 2.2) are sometimes included under the category of ultra-trace pollutants. Once again, a determination of each individual compound is necessary, with the most toxic compounds being those without chlorines in the 2,2′ positions. These are found in higher concentrations than PCDDs and PCDFs and can be separated as part of the extraction schemes to be discussed later. There may be occasions when compounds already discussed as trace pollutants (see Chapter 4) may need to be monitored at ultra-trace levels. Examples could include compounds which are currently being investigated as endocrine disruptors. I will, however, restrict the following discussion to PCDDs and PCDFs as similar techniques can be applied to the other groups of compounds.

DQ 8.2

The solubility of 2,3,7,8-tetrachlorodibenzo-*p*-dioxin in water is 0.019 $\mu g\ l^{-1}$ at 25°C. This is a solid with a vapour pressure at normal temperatures of 6.2×10^{-7} Pa (atmospheric pressure is approximately 10^5 Pa). Using the considerations in the earlier sections, what can you deduce of its environmental behaviour? What would be the most suitable samples to take for environmental monitoring?

Answer

Using a relative molecular mass of 322, the solubility is equivalent to 5.9×10^{-5} $\mu mol\ l^{-1}$. *If you refer to Figure 2.4 earlier, which correlates solubility in water with bioconcentration factor, you will find that the factor will be extremely high. It is in fact off the scale in Figure 2.4. The compound is also likely to accumulate in sediments (see Section 2.3). The low vapour pressure indicates that the dioxin in the atmosphere will be predominantly in the solid state, and particulate analysis would be of major importance.*

Due to the low solubility in water and high bioconcentration factor, many of the early investigations were of dioxin concentrations in sediments and living organisms. Few investigations were of concentrations in natural water samples as these would have been expected to be at or below the lower detection limits.

An additional property of PCDDs and PCDFs is their strong binding ability to organic material in soils. Relatively high concentrations may be found in

Table 8.1 Typical concentrations of PCDDs and PCDFs

Congener group	Rural soils (ng kg^{-1})	Urban air (pg m^{-3})	Sewage sludge (ng kg^{-1})	Human fatty tissue (ng kg^{-1})
Tetrachlorodibenzo-*p*-dioxins	3–8	< 0.02–6.5	< 0.01–0.37	3–10
Tetrachlorodibenzofurans	5–30	< 0.02–18.7	< 0.01–0.90	3–9

contaminated soils due to the binding preventing dispersion of the material. The soil also prevents photolytic degradation which may occur when the compounds are exposed to sunlight.

Typical concentrations of PCDDs and PCDFs are shown in Table 8.1. The term 'congener group' is used to describe compounds with similar structural features, such as having similar numbers of chlorines or having a common substitution pattern regardless of the total number of chlorines present.

Compare these concentrations with typical values of other organic pollutants in soil and sludges (Sections 5.4 and 5.7), the atmosphere (Section 7.1), and living organisms (Section 5.3), and you will find that the PCDD and PCDF levels are lower by a factor of 1000 or more.

The separation and subsequent determination of 75 PCDDs and 135 PCDFs is quite a formidable task, even disregarding the low concentrations involved and possible interference by large numbers of other components in the sample. Most investigations restrict the analysis to compounds with four or more chlorines since the highest toxicity is where there are four to six chlorines per molecule with substitution in the 2,3,7,8-positions (i.e. members of the 2,3,7,8-congener group with four to six chlorines). Results can be quoted as individual concentrations or as a total 2,3,7,8-tetrachlorodibenzo-*p*-dioxin toxic equivalent (TEQ) concentration. This is calculated by weighting the concentration of each individual component according to their relative toxicities, with the scaling factor for 2,3,7,8-tetrachlorodibenzo-*p*-dioxin being 1.

SAQ 8.1

In light of the low concentrations involved, suggest why we have to measure each individual PCDD or PCDF rather than the total PCDD or PCDF content.

8.2 Analytical Methods

8.2.1 General Considerations

Let us start with our existing knowledge of the analysis of organic micropollutants.

DQ 8.3

What were the main stages of the analytical determination?

Answer

(i) Extraction of the analyte.

(ii) Separation from interfering compounds by chromatography.

(iii) Concentration.

(iv) Analytical separation and determination by gas chromatography, for chlorinated compounds by using either an electron capture detector or mass spectrometric detector.

You should revise Section 4.2 if you have not remembered these steps.

We must now consider modifications to the method to analyse ng l^{-1} rather than $\mu g\ l^{-1}$ concentrations. There would appear to be two routes, as follows:

1. Increasing the overall concentration factor in the pretreatment stages.
2. Increasing the detection sensitivity.

There may be room for a small improvement in overall sensitivity by method 1 but this will be insufficient on its own. Note, for instance, the final extract volume (1 ml) given earlier in the DDT analytical method in Section 4.2. There could be at least a 10-fold reduction in this volume, with a corresponding increase in sensitivity. The possibility of increasing the sample size is limited by physical considerations of handling large samples under clean conditions. A disadvantage of this method is that any impurities carried through the extraction scheme will also tend to concentrate. Method 2 could lead to a more substantial increase in overall sensitivity, particularly if selective detection of the desired analyte is included.

8.2.2 Factors Affecting Detection Sensitivity

The determining factor for this is often the 'random' baseline fluctuations in the chromatogram due to unresolved minor peaks. These peaks could be due to components not removed in the sample pretreatment.

DQ 8.4

What approaches do you think could be used to minimize this effect?

Answer

1. ***Increasing the chromatographic resolution of the column.*** *This would decrease peak overlap and more readily allows minor peaks to be distinguished from the true baseline. High-resolution capillary columns are essential for ultra-trace analysis.*

2. ***Increasing pretreatment to increase the removal of minor components***. *This will increase analytical time (which can already be of the order of 24 h). Each additional step increases the possibility of sample loss or contamination.*
3. ***Changing to a more selective detector which would not respond to the minor peaks***. *Most current investigations use some form of mass spectrometric detection. This may allow a simplified pretreatment but can increase, sometimes almost prohibitively, the cost of instrumentation and hence decrease the number of laboratories which can be equipped to perform the analysis.*

The effect of the three approaches is shown in Figure 8.2. The wide variety of analytical methods found in the literature result from different emphases being

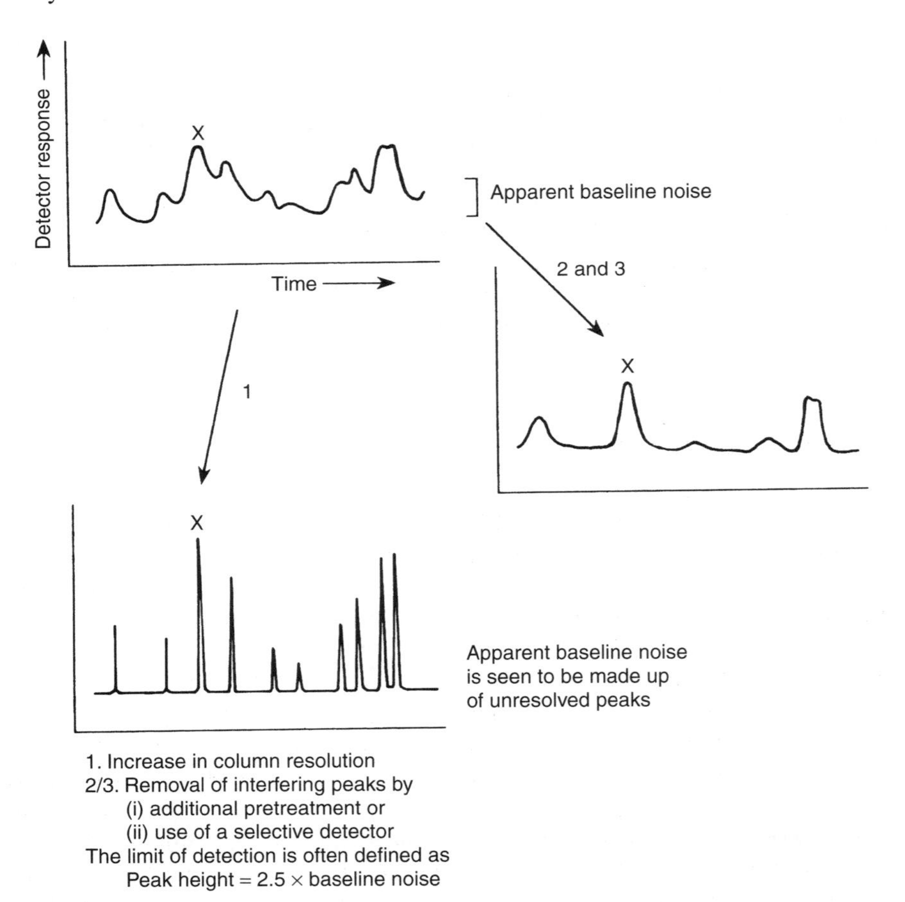

Figure 8.2 Methods of increasing detection sensitivity.

placed on the last two methods. In order to understand this, we now need to examine in detail how mass spectrometers may be used as selective GC detectors.

8.2.3 Mass Spectrometric Detection

You should have already come across mass spectrometry as a method of identifying organic compounds and in Chapter 4 we mentioned its use as a GC detector.

DQ 8.5

Briefly describe this technique (MS) and how it is used to identify organic compounds.

Answer

The compound is ionized under high vacuum, often using electron impact. In the process, the molecules fragment. The ions produced are focused into a beam, accelerated and then separated according to their masses (or more precisely their mass/charge (or m/z*) ratios).*

High-resolution (double-focusing) mass spectrometers separate the ions by using both magnetic and electrostatic fields (Figure 8.3). Low-resolution spectrometers, which are commonly found in bench-top gas chromatograph–mass spectrometer set-ups, tend to use electrostatic quadrupole separation or ion-trap techniques (Figure 8.4).

The peak with the highest mass/charge ratio is usually (but not always) from the unfragmented ion. This can be used to confirm the relative molecular mass of the compound. The fragmentation pattern (Figure 8.5) can given an indication of the chemical groups in the molecule. Under favourable circumstances, the

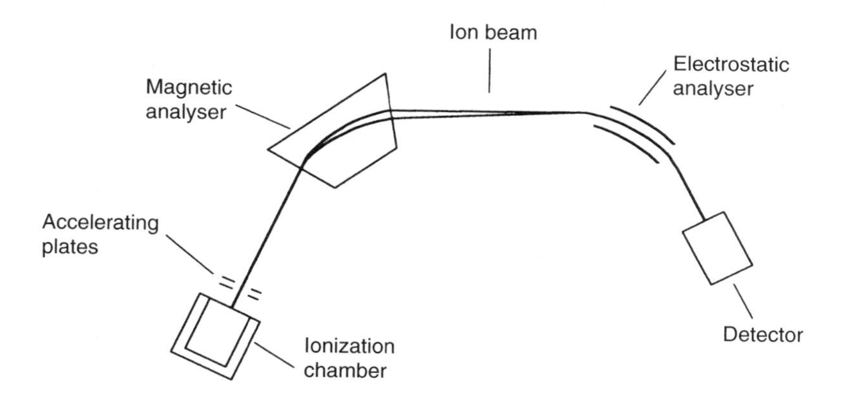

Figure 8.3 Schematic of a high-resolution mass spectrometer.

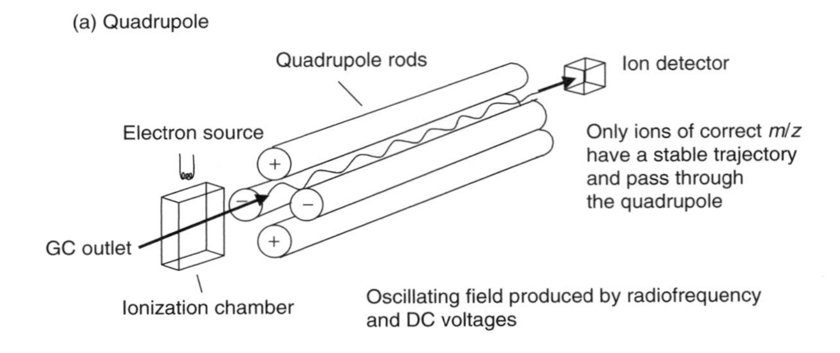

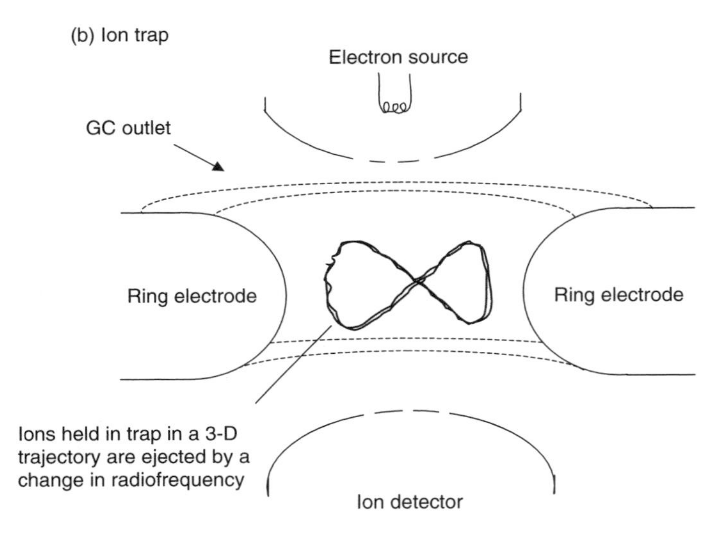

Figure 8.4 Examples of bench-top mass spectrometric detectors: (a) quadrupole; (b) ion trap.

molecular structure can be determined from this. However, a simpler method of identification is to compare the spectrum with that from a pure sample, or from a reference library.

In order to use the mass spectrometer as a universal GC detector, the total-ion current is monitored. A typical chromatogram is shown in Figure 8.6. When dealing with simple mixtures, the chromatographic peaks can then be identified, and their purity confirmed by the production of a complete mass spectrum for each peak or part of a peak. Even with this simple use of a mass spectrometer, you can see how much data can be generated and why the widespread use of GC–MS had to await development of cheap computer data storage!

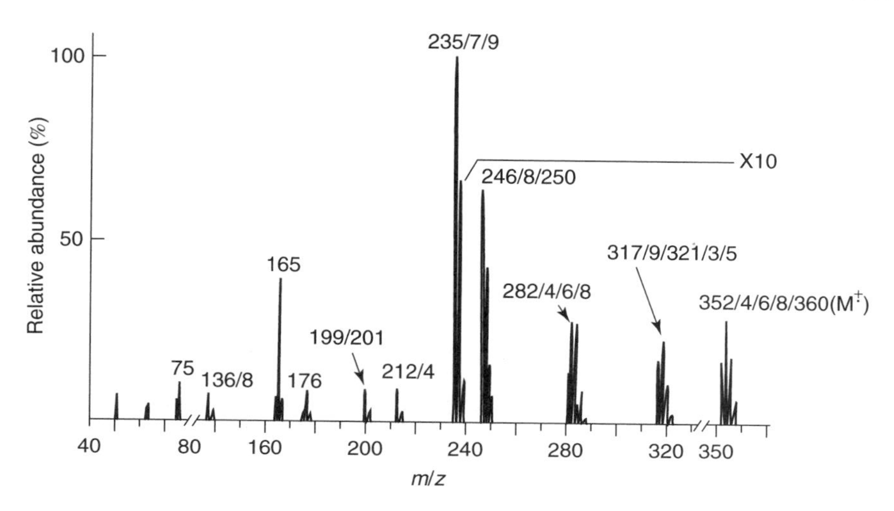

Figure 8.5 The mass spectrum of *p,p′*-DDT, displaying a typical fragmentation pattern. From Barker, J., *Mass Spectrometry*, 2nd Edn, ACOL Series, © University of Greenwich, 1999. Reprinted by permission of University of Greenwich.

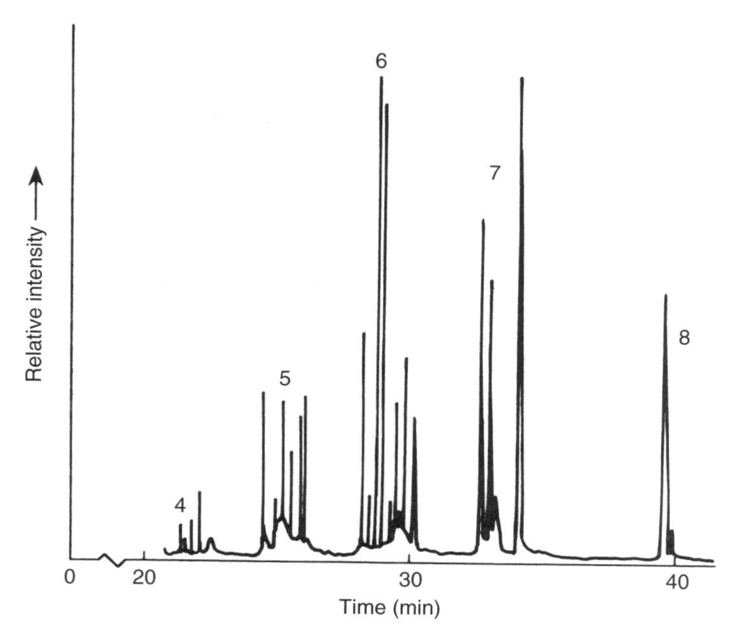

Figure 8.6 Total-ion chromatogram of a dioxin mixture; the numbers assigned to the various groups of peaks indicate the number of chlorine atoms in the molecules.

One of the problems of the analysis of PCDD/PCDF mixtures is that different species can give identical fragmentation patterns and it is difficult to identify a species on the basis of its mass spectrum alone.

The above application of GC–MS is useful for preliminary chromatographic surveys but still does not use the capability of the spectrometer as a selective detector. The simplest way to do this is to monitor a single ion, with this usually being the molecular ion of the compound, i.e. the ion with the same molecular mass as the parent molecule. There is an increase in sensitivity in comparison to total ion current detection since the detector spends all of its time monitoring one ion rather than scanning the complete range. The chromatogram produced contains fewer peaks than a total ion current chromatogram, but in a mixture such as a dioxin extract, the chromatogram may still be complex. If the detector is set at *m/z* 322, for instance, all 22 tetrachlorodibenzo-*p*-dioxin isomers will be detected as well as ions from other compounds which coincidentally have the same *m/z* value.

Potential interferences in the chromatogram can be detected if fragments are monitored at two or more mass/change ratios. This is known as selected ion monitoring. When applied to dioxin analysis, the technique makes use of naturally occurring chlorine being found as an approximate 3:1 mixture of ^{35}Cl and ^{37}Cl isotopes. Any molecular fragment containing one chlorine atom will be able to be detected at two mass/charge ratios separated by 2 atomic mass units, corresponding to the ions containing ^{35}Cl and ^{37}Cl. Their intensities should be in the ratio 3:1. If the fragment is not detected at both *m/z* values, then you have been wrong in your assumption that the fragment contains chlorine. If the relative intensities are not 3:1, and you are certain that there is just one chlorine in the fragment, then this would suggest there is interference from a second ion which coincidentally has an identical *m/z* to one of the ions.

If the fragment contains more than one chlorine, the pattern will become more complex, but still predictable and easily recognizable with experience. The relative intensities of ions containing between one and four chlorines is shown in Figure 8.7.

DQ 8.6

Which *m/z* values could be used to detect the unfragmented 2,3,7,8-tetrachlorodibenzo-*p*-dioxin ion?

Answer

The highest-intensity ions will be:

$$^{12}C_{12}{}^{1}H_{4}{}^{16}O_{2}{}^{35}Cl_{4} = 320$$

$$^{12}C_{12}{}^{1}H_{4}{}^{16}O_{2}{}^{35}Cl_{3}{}^{37}Cl = 322$$

$$^{12}C_{12}{}^{1}H_{4}{}^{16}O_{2}{}^{35}Cl_{2}{}^{37}Cl_{2} = 324$$

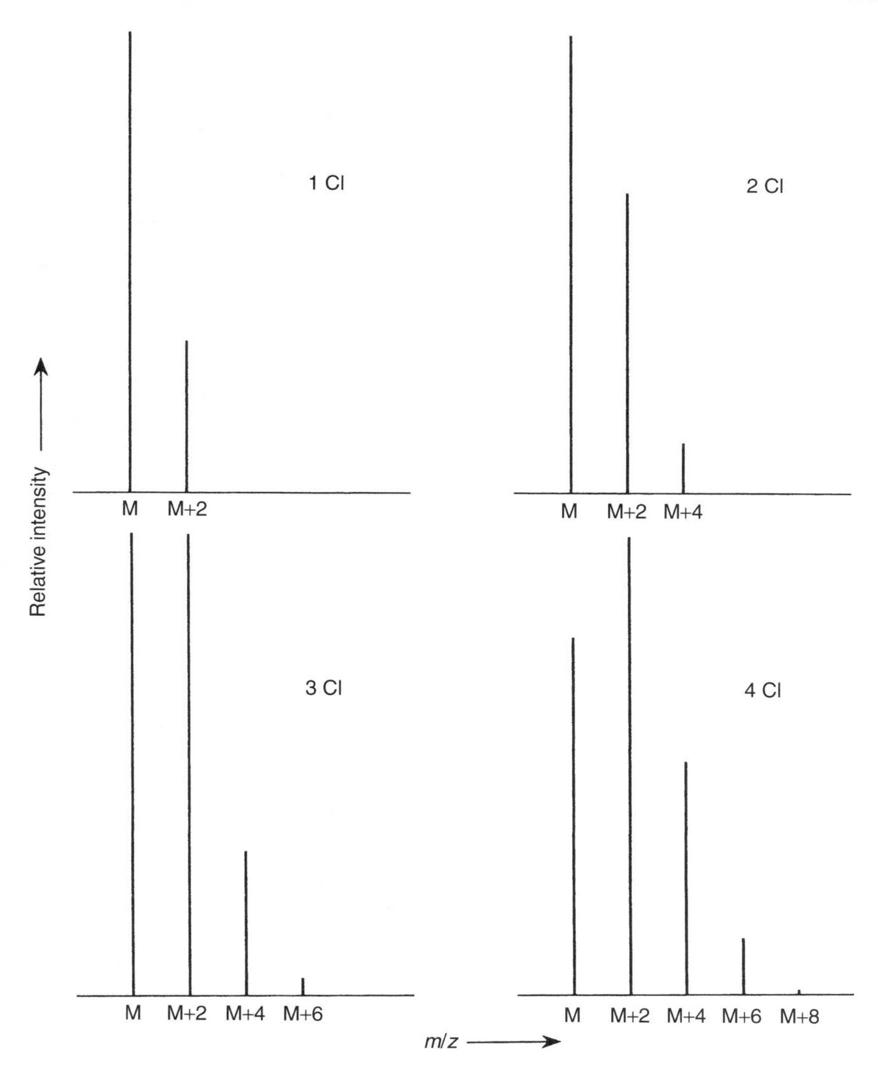

Figure 8.7 Relative intensities of ions containing more than one chlorine atom; M represents an ion containing entirely ^{35}Cl.

$$^{12}C_{12}\,^{1}H_{4}\,^{16}O_{2}\,^{35}Cl\,^{37}Cl_{3} = 326$$

$$^{12}C_{12}\,^{1}H_{4}\,^{16}O_{2}\,^{37}Cl_{4} = 328$$

There will also be a number of lower-intensity peaks due to ions containing one or more ^{13}C *atoms, rather than all of the carbon atoms being* ^{12}C *(the natural abundance of* ^{13}C *is 1.08%).*

Statistical considerations show that the *m/z* 320 and 322 ions are the most abundant (relative intensities 77:100) and, in practice, it is just these ions which are normally monitored.

You should note that, although the existence of chlorine isotopes is a considerable advantage for the identification of chlorine-containing fragments, it does increase the possibility of other species being detected at any chosen *m/z*. Consider *p,p*′-DDE (see Figure 2.6 earlier). The most abundant molecular ion has *m/z* 318 but there are lower-intensity peaks at 316, 320, 322 and 324. Two of these ions will potentially interfere with tetrachlorodibenzo-*p*-dioxin determinations at *m/z* 320 and 322.

DQ 8.7

As described so far, it would appear that selected ion monitoring possesses a major problem for the analysis of dioxin mixtures. What is this?

Answer

If the detection is set at, say, m/z *320 and 322, then only the 22 tetrachlorinated isomers will be detected, and not other polychlorinated dioxins. The detector could be set to include additional mass/charge ratios corresponding to the other species, but a compromise would have to be reached as the overall sensitivity decreases with an increase in the number of ions detected.*

The problem can be overcome by changing the ions monitored during the course of the elution. This, of course, leads to the requirement for the chromatographic column to separate the mixture into isomer groups. Most columns used for dioxin analysis are able to achieve this group separation but this is usually at the cost of incomplete resolution of some individual isomers. You may wish to look back at Figure 8.6 to check this statement for a typical chromatographic separation, and to determine possible ion detection sequences which could be used for selected ion monitoring.

We have now discussed both parts of the most common detection method used in dioxin analysis, namely:

- Selected ion detection at two or more *m/z* values.
- Detection of different isomer groups by change of the *m/z*s monitored throughout the chromatogram.

There are, however, a number of other approaches to selective detection. Two of these will be discussed below.

8.2.3.1 Other Mass Spectrometric Methods

Although some ions may appear to have identical masses on low-resolution spectrometers, they may often be differentiated by using high-resolution instruments. For example, the following ions (molecular ions or fragments) would be detected at *m/z* 322 on a low-resolution instrument:

Ion	*Accurate mass*
Tetrachlorinated dioxins	321.8936
DDE	321.9292
DDT	321.9219

These may be selectively detected by high-resolution mass spectrometry due to slight differences in their accurate masses. Although this is an obvious advantage as the pretreatment may be reduced, the majority of literature methods still use low-resolution spectrometry due to its lower cost and wider availability.

A further development is tandem MS/MS where a single ion (say *m/z* 322) is subjected to a second fragmentation to confirm the identity of the ion and to permit determination without interference. Many of the original investigations used connected quadrupole spectrometers or a double-focusing and quadrupole combination. The detection limits were generally lower than conventional GC–MS. Initial hopes in the development of the method were that such a technique would completely remove the necessity for a pretreatment stage, even for dioxins in complex sample matrices, and hence produce a considerable analytical time-saving. Pretreatment was, however, still found to be necessary for some samples.

Bench-top GC–ion trap mass spectrometers can be operated in the MS/MS mode, in which ions are subjected to a second fragmentation within the trap. The *m/z* 322 tetrachlorinated dioxin ions fragment to form *m/z* 259 ions which can then be determined free from the interferences discussed above. This can be used for rapid routine screening of dioxins, although the instruments are significantly less sensitive than high-resolution instruments.

8.2.4 Quantification

This is normally performed by the addition of known amounts of standards to the sample before extraction (see Section 4.2 earlier). The method will compensate for sample losses in the clean-up stage, assuming that the losses of the standard are identical to those of the analyte, and will also ensure the determination is independent of any variations in the sensitivity of the spectrometer. Fully substituted ^{13}C isotopically labelled compounds are often used.

Ideally, one ^{13}C standard should be added for each compound to be determined. This is, however, not generally practicable. It is common practice to use just one standard for each isomer group, and this would normally be the compound containing the 2,3,7,8-substitution pattern. Isotopically labelled derivatives are available for all 17 PCDDs and PCDFs which contain this substitution pattern.

Concentrations of non-2,3,7,8 isomers can be calculated if their responses with respect to the 2,3,7,8 compound can be determined by using reference samples of pure material. These may not always be available. In general, however, other compounds are not determined individually, with isomer group concentrations (e.g. total tetrachlorinated PCDDs) generally being considered sufficient. Look back at Table 8.1 to find an example of their use. The group concentrations are determined by using an average response factor calculated from as many individual isomer response factors as are available.

8.2.5 Quality Control

A second standard is often added immediately before injection into the GC–MS system. The standard can be a ^{13}C or ^{37}Cl labelled compound. This is for quality control purposes. It allows determination of the recovery of the dioxin over the clean-up stage. A low recovery would give rise to concern over the accuracy of the final results. The second standard can also be used to provide an estimate of the sensitivity of the detector, which may vary over a period of time. This is an important control feature as most determinations involve operating the instrument close to the limit of detection.

SAQ 8.2

The analytical technique for PCDDs and PCDFs from solids includes extraction, followed by clean-up and concentration of the extract, and then GC–MS analysis. Why is an isotopic standard better than, say, a compound which is structurally similar to PCDDs and PCDFs and is not found in the analytical mixture?

SAQ 8.3

Using GC–MS with selected ion monitoring, how would you set about confirming the identity of a low-intensity chromatographic peak as a particular dioxin?

8.3 A Typical Analytical Scheme

The pretreatment is summarized in Figure 8.8 and the subsequent chromatographic separation in Table 8.2. The scheme is an example of a method where the emphasis is placed on sample clean-up and separation combined with low-resolution mass spectrometry, rather than relying on high-resolution MS techniques. I wish to use this scheme as an exercise to test your understanding of the principles of sample pretreatment and the subsequent analytical determination.

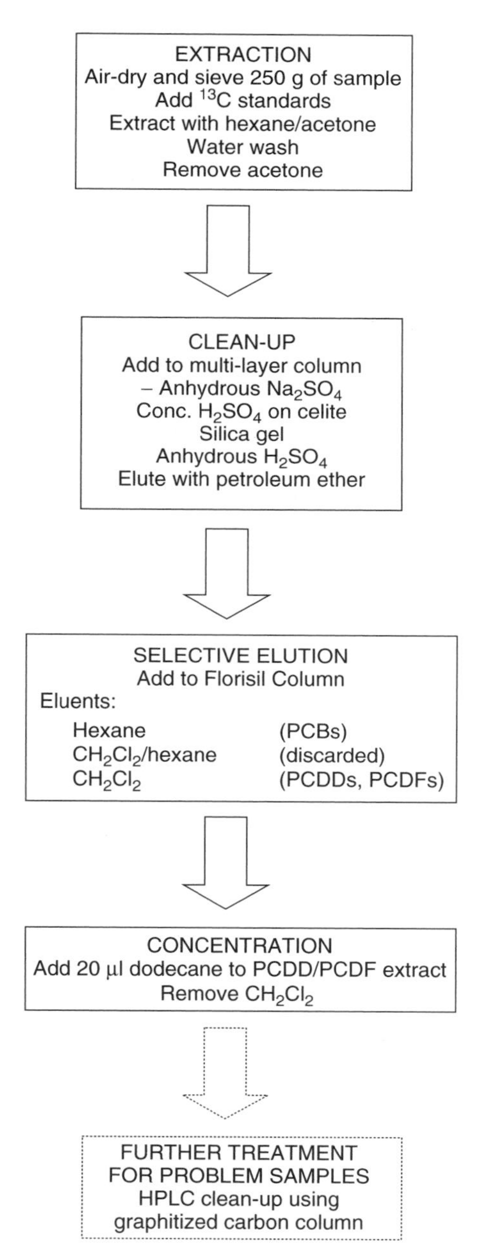

Figure 8.8 A typical pretreatment scheme for soil. From 'Determination of polychlorinated biphenyls, polychlorinated dibenzo-*p*-dioxins and polychlorinated dibenzofurans in UK soils', Technical Report, Her Majesty's Inspectorate of Pollution, HMSO, London, 1989.

Table 8.2 Chromatographic conditions used in the analytical determination of soils[a] (cf. Figure 8.8)

Tetra- and pentachlordibenzo-*p*-dioxins and furans:

- 5 m × 0.2 mm i.d. BP-5 (medium polarity) capillary column, plus
  50 m × 0.2 mm i.d. RSL-950 or CP-Sil 88 (highly polar) capillary column
- Oven temperature gradient, 170–240°C
- Elution time, 20 min

Hexa-, hepta- and octachlorodibenzo-*p*-dioxins and furans:

- 50 m × 0.2 mm i.d. BP-5 capillary column
- Oven temperature gradient, 170–290°C
- Elution time, 20 min

[a] 'Determination of polychlorinated biphenyls, polychlorinated dibenzo-*p*-dioxins and polychlorinated dibenzofurans in UK soils', Technical Report, Her Majesty's Inspectorate of Pollution, HMSO, London, 1989.

This will form a suitable conclusion to this open learning material on environmental analysis.

8.3.1 Pretreatment

First of all, let us compare the scheme with the analysis of DDT as discussed earlier in Section 4.2. You should note the overall similarity of the individual stages.

DQ 8.8

What are the major differences between the two analytical schemes?

Answer

1. ***The initial chromatographic clean-up for the dioxin analysis uses more than one stationary phase, thus reflecting the complexity of the extract.*** *The number and type of phases used for dioxin extract clean-up vary considerably between literature methods. Acidic and basic silica and alumina columns are in common use. Note here the use of a multi-layer column rather than individual columns, hence saving analytical time and minimizing the possibility of sample contamination or loss.*
2. ***Additional pretreatment for 'problem' dioxin samples.*** *HPLC separation using graphitized carbon is used. This form of carbon has been found to be highly selective towards planar molecules. It is a straightforward, although time-consuming, operation to determine which sample types need further pretreatment, by comparison of the GC–MS chromatogram of one sample with and without the additional step.*

Note, for both the steps given above, the necessity to minimize pretreatment time while still maintaining the efficiency of the clean-up procedure.

DQ 8.9

Why do you think there is a change of solvent composition between extraction and clean-up for the dioxin analysis?

Answer

The extraction stage uses a hexane/acetone mixture. Acetone is often used as a solvent modifier in extractions from solids to increase the polarity of the solvent and to assist in the penetration of the solvent into the samples (see Section 5.3 earlier). The first clean-up procedure involves application of the extract on to a chromatography column and subsequent elution of the non-polar components with a non-polar solvent (petroleum ether). The presence of a polar solvent in the extract would lower the efficiency of the chromatographic separation.

In both of the DDT and dioxin analytical procedures, a second column is required to separate chlorinated species (pesticides, PCBs, etc.).

DQ 8.10

Why do you think these components were not removed by the first column?

Answer

The chlorinated compounds have similar chemical structures. They are all neutral, non-polar, high-molecular-mass compounds and will have similar chromatographic retention properties. The first column in any clean-up is generally to remove interference from compounds with widely different chromatographic properties. The non-polar eluent used will elute the chlorinated species together. Separation of these closely related species will require a second and more selective column with sequential elution of the compounds by a series of solvents of increasing polarity.

The necessity of ensuring purity of solvents and cleanliness of apparatus has been discussed earlier in Sections 2.9 and 4.2 and needs to be re-emphasized here. All batches of solvents and reagents need to be frequently checked to confirm lack of contamination. Pesticide-free or distilled-in-glass grade solvents should be used. Extreme care should be taken with respect to known sources of dioxins. Cigarette smoke and ash can contaminate the laboratory. Extraction thimbles used for solids can be a second source, with dioxins potentially being formed by the bleaching process during their manufacture. The thimbles should be pre-extracted with solvent prior to use in the analysis.

8.3.2 Gas Chromatography

The chromatography column has both to separate the components of the mixture and be compatible with mass spectroscopic detection.

DQ 8.11

From your knowledge of the analytical problem, what can you say about the type of column required?

Answer

1. *First of all, in order to resolve the total number of components and to interface with the mass spectrometer, narrow-bore capillary columns are necessary. A programmed temperature gradient will optimize the separations.*
2. *The stationary phase would have to be compatible with high-temperature operation in order to elute the lower-vapour-pressure compounds. A silicone stationary phase would be the obvious choice.*
3. *With a mixture of compounds of varying polarity (according to the degree of substitution and substitution pattern), a medium-polarity stationary phase would be a good first try.*
4. *For mass spectroscopic detection, it would be advantageous for the column to group the eluted compounds into isomer groups. This aids peak identification, as well as allowing a fairly simple ion sequence to be used for detection.*

The separation of all 210 dioxins and furans and their division into separate isomer groups are exacting demands for a single capillary column. You may not be surprised to find that two columns and multi-dimensional GC/GC were used in the analytical scheme (see Table 8.2).

SAQ 8.4

Imagine that you were about to analyse a large number of samples by a method such as that described in Section 8.3. What features would you include in your scheme to ensure analytical quality throughout the programme?

SAQ 8.5

There are at least two areas of uncertainty in the analytical procedures for the determination of dioxins in solid samples as exemplified above. What are these?

Summary

Some species (e.g. dioxins and related compounds) have such a great ability to bioaccumulate and such a high degree of toxicity that monitoring their presence at ng l^{-1} or ng kg^{-1} concentrations is necessary. The analyses not only require highly sensitive and selective instrumentation but also a large degree of analytical skill and expertise. Gas chromatography–mass spectrometry is most often used. This technique has been discussed, along with the necessary clean-up and concentration stages.

Responses to Self-Assessment Questions

Chapter 1

Response 1.1

1. You could write many pages on this first question. The particular problem mentioned here is the disproportionate proportion of the world's resources used in the developed world, together with a similar proportion of the waste produced and pollution of the earth.
2. Although volcanic emissions are natural phenomena, at intervals they put into the atmosphere large amounts of gases, vapour and dust, and could be considered a natural source of pollution.
3. The production of methane by cows is again a natural phenomenon. However, since the total cattle population on the earth is largely controlled by mankind, then so is the quantity of methane emitted to the atmosphere from this source.
4. This will directly lead to an increase in the concentration of nitrate above the naturally occurring levels in water supplies surrounding the farms. There may, however, be other consequences. Since all of the species in the nitrogen cycle (see Figure 1.1) are linked, changes in nitrate concentrations may lead to changes in concentrations of other species in the cycle, thus leading to further pollution problems (see Table 1.1).

These are all examples of current concern. As countries become developed, more of the earth's resources are used. The volcano of Mount Etna in Italy is thought to be a significant contributor to atmospheric mercury concentrations in Europe. The rising atmospheric methane concentration has, in part, been attributed to increasing cattle populations.

Concern has been expressed over the effect of increased nitrate fertilizer usage in the production of nitrous oxide, a greenhouse gas and potential ozone-layer depleter.

Response 1.2

Analysis of the discharge before dispersal into the river will monitor the pollutant being discharged, but will only be related to the final concentration in the river when used in conjunction with river flow data.

Analysis of the river sufficiently downstream from the discharge point to allow for dispersal will give a direct measurement of the concentration in the river, but there will be some uncertainty as to the source of the pollution.

Analysis of the pollutant in living organisms found in the river will give a direct indication of the environmental problem, but unless the organisms are sampled close to the discharge point, the analytical results would be difficult to relate to individual discharges.

The discharge composition was included to encourage you to extend your ideas to consider what extra analytical information would be useful. Quality control of the starting materials would reduce the concentration of the contaminants in the discharge, as would process control to minimize the manufacture of side products.

National legislation may in fact specify the requirements of monitoring such as detailed in this response.

Chapter 2

Response 2.1

The first prerequisite is that the compound must be in a form which allows it to become widely dispersed. Dispersal may be via the atmosphere or the hydrosphere. The properties which would affect this dispersal include volatility, solubility in water, and if the compound is a solid, its particle size.

The compound must have a high resistance to degradation within the atmosphere and hydrosphere and to metabolism (chemical reaction) within organisms. If the compound rapidly degraded, there would be little possibility of toxic concentrations building up. Beware, however, of the possibility that the degradation products themselves might pose similar or worse environmental problems.

If the pollution problem is related directly to the effect of the material on living organisms, rather than on physical structures, then the compound must have an ability to reconcentrate within organisms. This will be described in detail in the following sections.

Finally, if the compound is seen as a pollution problem, it must have some deleterious effect. However, this need not always be the case.

Response 2.2

DDT is a high-molecular-mass neutral organic compound. As with all similar compounds, it would be expected to have a significant vapour pressure. The following illustration indicates some possible dispersal routes – drift of the spray from the field of application, run-off into watercourses, and also volatilization.

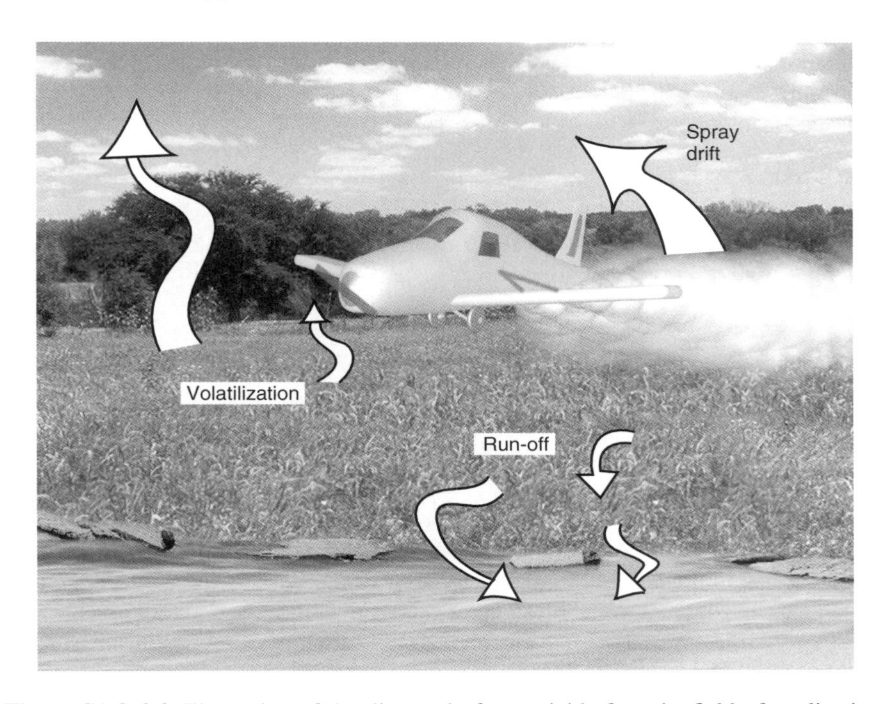

Figure SAQ 2.2 Illustration of the dispersal of a pesticide from its field of application.

Response 2.3

Both types of pollutant may be dispersed through the atmosphere, as well as in watercourses. Atmospheric dispersal of metals is usually as particulates (in the form of metal salts), whereas organics are found both in the particulate form and in the vapour state.

Deposition may occur on to land or into water courses.

The solubility of the organic compounds in water is often low, but this can lead to a large bioconcentration in organisms living in the water. The solubility of metals is very dependent on the chemical compounds involved, but for most species will increase with a decrease in pH. There is normally no pH effect on the solubility of neutral organic compounds. The bioconcentration of metals is very dependent on the element being considered; however, for some metals, such as cadmium, this can be extremely high (see Table 2.1).

Both types of pollutant will concentrate in sediments. The detailed mechanism for this is quite different in the two cases. Reducing conditions would, for instance, increase the deposition of lead, whereas these would have no effect on neutral organic compounds.

Entry into the food chain in an aquatic environment is in both cases by bottom-dwelling fish and filter-feeders. Once again, the detailed mechanisms are different in the two cases and this may lead to concentration in different organs in the body.

If you had said that there were some similarities in the dispersal and reconcentration you would have been correct, but there are many differences in the detailed mechanisms.

Response 2.4

The metal ions would be likely to concentrate in the sediment on the sea bed and be ingested by filter-feeders. A likely critical group would be the local people who consume large quantities of the sea-food, perhaps the families of the fishermen themselves.

A second likely path could exist. Some of the sediment may be washed up on the shore. This may dry out and be blown into the atmosphere, or it may be propelled into the atmosphere via sea spray.

The critical group could then be people who spend a large proportion of their time on the sea-shore, with the metal ions entering their bodies through inhalation.

Response 2.5

In a highly polluted area, there is the possibility that contamination may occur even at the sampling stage. This is particularly the case if the same apparatus is being used for a number of samples without thorough cleaning. A suitable sample container must be used to prevent contamination or loss during storage and transport. We have already discussed (see Section 2.6) that this should be glass rather than plastic for organic compounds. When working in the laboratory, there is the possibility of contamination by solvents in the laboratory atmosphere. At the instrumental analysis stage, there is the possibility of cross-contamination by consecutive samples. A suitable quality control procedure would be to analyse blank samples which have been introduced at the stages when contamination is likely – at the sampling stage, before transportation, before storage in the laboratory, and immediately prior to the instrumental analysis (see Table 2.3). Not only would any contamination be identified but also the point at which it was introduced.

Chapter 3

Response 3.1

Inorganic ions would be expected to increase in concentration further downstream due to evaporation of water, as well as from continual weathering of rocks.

The total loading of suspended solids can increase downstream as more solid material accumulates in the stream. Often, however, you will find large amounts of deposition in slow-moving areas.

The buildup of organic compounds will depend on the rate at which they are being introduced into the river compared with the rate of their oxidation. The latter will, in turn, depend on the rate of re-oxygenation of the water. If this is rapid, then the organic content will decrease; if not, the oxygen concentration will decrease and the organic content will then build up.

Re-oxygenation (and hence the oxygen concentration) will be high when the river is fast flowing near its source. Further downstream, where the river is more placid, oxygen uptake will be slower. Some, or in the worst cases, all, of the oxygen may be consumed by reaction with organic material.

Response 3.2

The top layer is exposed to the atmosphere and sunlight. Photosynthesis can take place producing oxygen and may lead to a lowering of nutrient concentrations. The layer will also be oxygenated from the atmosphere. Chemical species which can exist in either oxidized or reduced forms will be oxidized.

The bottom layer is not in direct contact with the atmosphere, thus leading to lower dissolved oxygen levels. Photosynthesis will be reduced. There will, however, be more opportunity for interaction and exchange of chemicals with the sediment on the lake bed.

Response 3.3

Decisions will have to be made concerning the following:

- the analyses required
- the timing of the sampling (i.e. the sample programme)
- the number of samples to be taken
- the location of the sampling
- sample volumes and containers
- the method of sample storage

Ammonia is an alkaline gas which would very easily escape from the sample. The solution could be acidified, so converting the ammonia into the ammonium ion. This is less readily lost:

$$NH_3 + H^+ \rightleftharpoons NH_4^+$$

Glass or polyethylene storage containers can be used.

Chloroform is similarly volatile but it is not easy to think of a simple method of fixing it in solution. The only easy method to minimize volatilization losses is to completely fill the sample container, store at sub-ambient temperature, and to keep the storage time as short as possible. A glass container should be used. Polyethylene containers may contaminate the samples with compounds which interfere in the subsequent analysis.

Organic compounds are readily oxidized by micro-organisms using oxygen from the air. To minimize the biological activity, the container should be completely filled and stored at 4°C. A glass container should again be used.

Response 3.4

Dissolved oxygen is necessary to support animal life within the river. For an unpolluted river, this should be close to saturation. Any organic pollution would tend to diminish this value.

BOD measures the oxygen depleting potential within the river. A small quantity of material (a few mg l^{-1}) will generally be present in any river from decaying vegetation. However, once the BOD value rises above this level, there is the potential of substantially depleting the oxygen content of water. (Remember that saturated water has a concentration of 8.54 mg l^{-1} oxygen at 25°C and 1 atm pressure.)

Ammonia is a natural constituent of water, being formed by the decomposition of organic material. It never reaches high concentrations under normal conditions as it is quickly oxidized to nitrate. An increase in concentration would be a further indication of the poor oxygenation of the water. High concentrations of ammonia are also toxic towards fish and would present a pollution problem in their own right.

Response 3.5

Cations

Ion	Absorption spectrometry	Ion chromatography	Titration	Flame photometry	Ion-selective electrodes
Na^+		×		×	
K^+		×		×	
Mg^{2+}		×	×		
Ca^{2+}		×	×	×	
NH_4^+/NH_3	×		×		×

Note that I have included Ca^{2+} and Mg^{2+} in the titration column. The standard method for water hardness gives a measurement of total polyvalent ions (largely Ca^{2+} and Mg^{2+}). You should have noted that a second titration at $pH > 10$ will measure calcium, and hence, by difference, magnesium can be estimated.

Atomic absorption spectrometry was also mentioned as being often used for magnesium analysis. As we will see later in Chapter 4, the technique could also be used for the other metals.

Anions

Ion	Absorption spectrometry	Ion chromatography	Gravimetric analysis	Ion-selective electrodes
Chloride	×	×		
Fluoride	×	×		×
Nitrate	×	×		
Nitrite	×	×		
Phosphate	×	×		
Sulfate	Indirect	×	×	

Response 3.6

The choice should depend on the following factors:

- Precision of the technique – compare with precision required.
- Analytical time of the technique – compare with urgency of result.
- Instrument time required for the technique – have you sufficient instruments?
- Analyst's time required for the technique. This can often be significantly different from the instrument time.
- Time required to set up any instrumentation, or in preparation of reagents. This time becomes more significant with small numbers of samples.
- Throughput of the laboratory. There may be instruments available for rapid analysis in a high-throughput laboratory (see Section 3.4.1).
- Number of analytes to be determined – note that some methods can determine more than one analyte.
- Availability of equipment.
- Relative cost of instrumentation/labour.

All of the criteria except the last two will be quite independent of the country in which you work. It would be possible for an instrumental method to be favoured in one part of the world where labour costs are high, whereas more labour-intensive methods are favoured in other parts of the world, and where, perhaps, instrumentation is less readily available.

Chapter 4

Response 4.1

First of all, let us decide which are the volatile compounds:

toluene, methylene chloride, chloroform and benzene (1, 4, 5, 10)

These will be analysed by the purge-and-trap technique.

There are two phenols in the list:
2,4,6-trichlorophenol and phenol (3, 7)
These would need to be extracted under acid conditions. (Under basic conditions, they would be in the form of non-extractable salts.)

The other compounds would form the base–neutral group:
anthracene, 1,2-dichlorobenzene, naphthalene, hexachlorobenzene (2, 6, 8, 9)

Do not be too concerned if you do not have all ten correct, but I hope you were able to think your way through most of them.

Response 4.2

Solid-phase extraction can be used in the field. You could perhaps load the column with the sample by using a syringe. Microfibres, with care, can be immersed directly in water in the field. A third method, which you may have considered, is purge-and-trap. This has been occasionally used. This method would give a concentration averaged over the purge period rather than an instantaneous value.

Response 4.3

(a) The pesticides will probably be present at the lower end of the trace-level range. Even after extraction and clean-up, a large number of compounds may still be present in the sample. A narrow-bore capillary column would give the required high resolution and detection sensitivity. A medium-polarity silicone polymer column would be a good initial choice.
(b) The components of interest would probably be part of a much higher concentration of waste chemicals. A wide-bore capillary column would be better suited than a narrow-bore column since it has a higher sample loading capacity. It would also be more tolerant of any non-volatile impurities in the sample. The column could also be coupled directly to the purge-and-trap system which you would almost certainly have used in the sample preparation. A medium-polarity silicone polymer column would again be a good initial choice.
(c) There would be a strong likelihood of non-volatile residues in the oil extract. A wide-bore capillary column would be more tolerant of contamination than narrow-bore ones. A non-polar silicone polymer column would be a good initial choice.

Response 4.4

N-*methylcarbamates* are analysed by HPLC using post-column derivitization or UV detection. Although not specifically stated, you should be able to deduce that MS detection could also be used. The polarity of the molecules and their thermal lability makes this technique preferable to GC.

The techniques specifically mentioned for *atrazine* in the present chapter are immunoassay and HPLC. HPLC will produce a result specific to the compound. With the immunoassay there may be cross-reactivity from other triazine pesticides.

If individual *phenols* were not required, the analysis could be performed spectrometrically. For individual phenols, HPLC could be used. This may be after the formation of a fluorescent derivative. Once again, the polarity of the molecules can make GC analysis difficult.

Individual *PAHs* can be determined by HPLC with fluorescence detection. Their inherent fluorescent properties makes this an ideal method. GC may also be used. If individual PAHs were not required, they can be determined by immunoassay.

Malathion is not mentioned in this chapter, although you should by now realize that GC is in fact the first choice technique for all organics unless there are specific reasons (such as in the examples above) why other techniques are preferable.

Response 4.5

Loss of analyte at each stage is possible. This will be a particular problem if it is present in low concentration. For the determination of common pollutants, contamination of the sample may also occur. The problem increases with the number of stages of pretreatment, and the number of reagents involved. Common metal ions are universally found at low concentration in all reagent solids and traces of pesticides are common in organic solvents. Low-molecular-mass organic materials (e.g. solvents) are themselves commonly in use within laboratories. All materials in contact with the sample should be regarded as potential sources of contamination.

Had I restricted my question simply to metals, you might also have included the problem of different metal species in the sample. Unless you had converted the metal species completely to a single form, you would risk the loss of metal at each stage of the analysis according to the chemical behaviour of the particular species.

Response 4.6

The pretreatment in (i) ensures that all of the metal species in the sample will have been decomposed and so the *total metal* concentration will be determined.

Procedure (ii) will measure the free metal ions and loosely bound complexes. This measurement is known as the *ASV-labile metal* content.

Procedure (iii) will destroy easily oxidizable organic material. Subtraction of the results, (iii) – (ii), will measure the amount of *metal found in the complexes*. The latter is known as the *organically bound labile metal* content.

Free metal ions will be held by the resin in procedure (iv), as will the metal from loosely bound complexes. Thus procedure (iv) will then determine the amount of *metal bound in highly stable complexes*.

Procedure (v) will extract organic soluble complexes, and hence subtraction of the results, (i) – (v), will measure the *organic-soluble* content.

Chapter 5

Response 5.1

The deposition would be expected to be greatest at the side of the road and would decrease with distance. A site should be selected where samples could be taken close to the road, ideally where there is no intervening pavement. The site should be distant from any other potential sources of lead. Both sides of the road should be sampled to compensate for the effects of wind. Sampling (taking duplicate samples) should be more frequent close to the road, perhaps with sampling distances from the road in the ratio 1:2:4:8:, etc. Since the source of the lead is from the atmosphere, surface samples should be taken. Samples taken below the soil surface could give an indication of penetration into the soil. Control samples as similar as possible to the monitoring samples should be taken some distance from the roadway.

Response 5.2

The use of a number of solvents will increase the number of extracted components. This would then require either more comprehensive clean-up procedures prior to chromatographic analysis, and/or a higher degree of resolution in the chromatographic separation. You may also have to consider problems associated with additional solvent peaks in the chromatographic analysis and the trace impurities introduced with each solvent.

Response 5.3

Potassium and calcium can be analysed by flame photometry (see Section 3.4.2) and magnesium and the trace metals by atomic absorption spectrometry or other atomic spectrometric techniques (see Section 4.3.3). Orthophosphate is best determined by visible spectrometry after conversion to a blue-coloured phosphomolybdenum complex (see Section 3.4.1).

Response 5.4

The order of sensitivity is as follows:

1. – (b) Atmospheric exposure of residents for long periods, with the possibilities of uptake of contaminants from crops and ingestion of metals in soil by children.
2. – (a) The public would spend a more restricted amount of time here, though there would be still be the possibility of direct exposure to contaminants in soil and atmosphere
3. – (d) More limited exposure to the public; much of the site may be surface-covered in subsequent development.

4. – (c) The hard surface of the car park would form a barrier to soil contaminants. The time spent by any one individual in such an area will be limited.

Volatile components are a problem in all areas, with methane a particular problem when it can build up within buildings ((a) and (c)). Toxic metals are a problem where there may be direct intake by, for instance, playing children or where crops are grown for possible human consumption ((a) and (b)). If you are considering heavily polluted industrial sites, you should also consider any aggressive properties from dumped waste (e.g. caustic properties and excessive amounts of leachable ion such as chloride) which may have an adverse affect on buildings as well as the public.

Response 5.5

You could be dealing with samples with concentrations of the analyte from percentage concentrations down to $\mu g\ l^{-1}$ and these could be taken from highly contaminated locations. There are problems of contamination at the sampling site, during storage and transportation (including cross-contamination of samples if inappropriate bottles or bottle tops are used) and within the equipment (carry-over of the analyte between samples).

The overall composition of waste samples can vary widely and quite often the total composition is unknown. Matrix effects may vary widely and unpredictably from sample to sample. Great care is necessary to confirm that these analytical problems have been overcome.

Response 5.6

Water will be the easiest to sample (unless you have access to local fisheries) and will need less pretreatment to remove potential interferences. The concentrations determined will be lower than in the other samples, and often are little above the lower limits of routine detection. (You may recall the typical concentrations given for DDT in Figure 2.5 and the metal enrichment factors shown in Table 2.1.)

Sediment will need more pretreatment to remove potential interferences. Concentrations will vary greatly from site to site and even from sample to sample and so a large number of samples would need to be taken to obtain an average concentration. Seaweed will also need pretreatment and will only be found in specific locations. On the other hand, sediment and seaweed samples are ideal for investigating localized pollution. In each case, the effect of enrichment or bioaccumulation make it easier to detect the species since they will be present at higher concentrations than in the surrounding water.

Fish are not static and so it is difficult to relate concentrations found to specific locations. There may be large variations in concentration from specimen to specimen. However, shellfish are more static and measured concentrations may be more easily related to localized pollution.

Response 5.7

All of the solids discussed in this chapter have complex structures and there will be a range of binding sites for the analyte. Some of the analyte can be so very strongly bound that it may be unavailable for uptake by organisms. It is quite possible that with length of time the pollutant becomes more tightly bound within the solid. We have considered this in Sections 5.4 and 5.7 for the related problem of metal availability in solids.

Chapter 6

Response 6.1

The definition of the term for gas concentrations is very precise, referring to measurements made as volumes. An atmosphere containing 20 ppm sulfur dioxide would contain 20 μl of gas per litre of atmosphere. A complete statement of the unit should be parts per million (*volume/volume*).

When the term is applied to aqueous concentrations, it is often used interchangeably with mg l^{-1}. This would give the complete statement of the unit as parts per million (weight/volume)! Since 1 l of water containing little dissolved material has a mass of 1000 g, this would become:

1 ppm = 1 mg of analyte per 1000 g of water, i.e.

parts per million (*weight/weight*) for water samples.

Response 6.2

Atmosphere	Typical pollutant concentrations (volume/volume)
External atmospheres	parts per billion (ppb) – parts per million (ppm)
Internal atmospheres	parts per million
Exhausts or flue gases	parts per million – parts per hundred (%)

1. *Different Concentration Ranges.* The concentrations in the two atmospheres span a range of 10^6. It is not surprising that some methods are more readily applicable to the low concentrations and others to higher concentrations.
2. *Different Analytes.* Unless you are concerned with highly localized pollution, the number of gaseous pollutants which can build up to detectable levels in the external atmosphere is small. Although Figure 6.1 presents by no means a comprehensive list, it does give an indication of the type of compounds which may be present – simple inorganic gases, a few stable organic compounds, and a number of photochemically generated species. A greater diversity can build up in internal atmospheres, and in particular, many organic compounds. These will require different monitoring techniques.

3. *Concern over Human Health.* You might expect, since internal atmosphere monitoring is largely concerned with human health, that instantaneous concentration measurements (or short-term, time-averaged concentrations) will be important, alongside longer-term, time-averaged values. Longer-term averaged values often predominate for external atmospheres. Different methods may be needed for the two types of determination.
4. *Sampling Difficulties.* We have not discussed this point before, but you should realize that air currents are usually lower and more stable indoors than in the external environment. Representative samples may be easier to obtain in an internal environment. We will find that both the accuracy and the precision of at least one of the methods we will be discussing is lowered by strong air currents. When a new analytical technique is introduced, there is sometimes a progression of validation, first for internal atmospheres and only subsequently for external atmospheres.

Response 6.3

1. Hydrogen peroxide solution is readily available.
2. The reaction product may be estimated by volumetric titration (e.g. using sodium hydroxide), removing the necessity of spectrometers and well-equipped laboratories.
3. The method is, however, non-specific. Any atmospheric component which will dissolve to form a strong acid or can be oxidized to a strong acid will be included in the final analysis.

For ambient air monitoring, potential interferences are likely to be at lower concentrations than the sulfur dioxide but for other analyses (e.g. flue gases) this may not necessarily be the case.

Response 6.4

Passive sampling techniques will require longer sampling times than the corresponding active sampling techniques, since they rely on gas diffusion. The minimum sampling time is several hours even for internal atmospheres, and so would be of little use in short-term monitoring.

Active sampling techniques have greater flexibility. The sampling rate can be adjusted, within limits, according to the application, thus making both long- and short-term monitoring possible. However, if used for personal monitoring, the pumps necessary for active sampling can be inconvenient for the wearer, and most would prefer to be monitored by passive sampling techniques.

Response 6.5

Solvent extraction can use standard laboratory apparatus. It is, however, time-consuming and can use potentially hazardous solvents.

Thermal desorption methods need a specialized instrument but these minimize laboratory manipulation. The sensitivity can be higher than for solvent extraction since the whole sample is introduced into the chromatograph in a single desorption. Only a small fraction of the extract is injected into the chromatograph in the solvent extraction method. However, replicate determinations are not possible with thermal desorption. This is easily achievable by using solvent extraction.

Response 6.6

1. *Absorption Trains*. These can sample continuously over a 24 or 8 h period to obtain a time-averaged value from a single analysis. In addition, they are less expensive than single instruments, thus leading to the possibility of simultaneous sampling at different locations. Furthermore, they can also be used as reference methods for other techniques.
2. *Gas Chromatography*. This technique is most frequently used as a central analytical facility for personal and multiple site monitoring, although portable instruments are available for initial site investigations.
3. *Direct-reading Instruments*. These are used for continuous monitoring of atmospheres at a limited number of sites. In particular, they are employed in national monitoring networks. Their high expense would limit their use for more localized, extensive monitoring exercises.

Response 6.7

(a) An absorption train could be used for the analyses. If a large number of sites were involved, passive samples could be used as an alternative, but with reduced precision. A specific NO_2 analyser (e.g. chemiluminescence) would be more appropriate for continuous single-site analysis.

(b) The most reliable method would be by sampling the atmosphere using adsorption tubes with subsequent gas chromatographic analysis. Gas detector tubes are available for common solvents, which would give almost instantaneous determinations, but care would have to be taken over possible interferences from other solvents which may be present in the laboratory.

(c) A personal monitor responding to carbon monoxide would provide protection. If a data logger is included, this would provide stored information for later, more detailed scrutiny. Continuous monitors (e.g. a non-dispersive infrared spectrometer) could also be located in the most hazardous areas.

Response 6.8

There is a choice of technique for most gas analyses. Considerations in the choice have to include 'ease of use in the field ' as well as other criteria (precision, cost, availability of equipment, etc.) which have already been discussed for laboratory analyses (see SAQ 3.6). This is illustrated in the following figure:

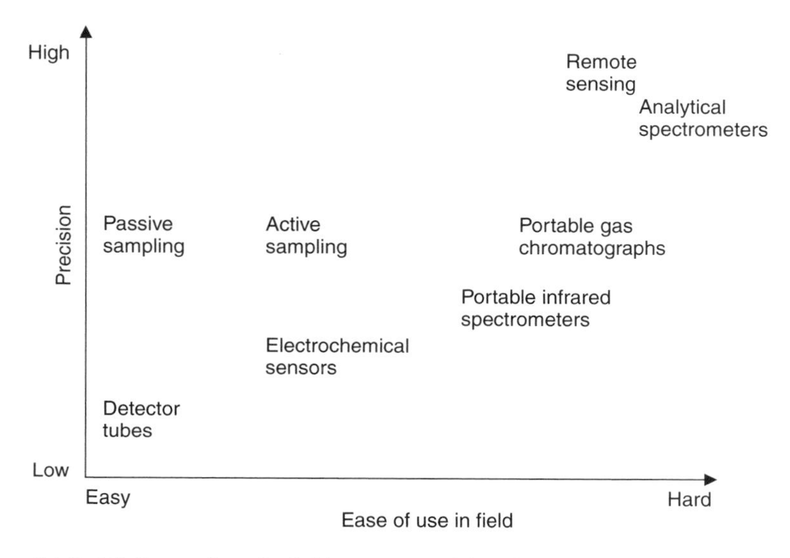

Figure SAQ 6.8 Ease of use in field versus precision for various gas analysis systems.

Chapter 7

Response 7.1

The most relevant sampling would be by using a personal sampler with the filter holder attached to the technician's lapel. However, this should be backed up with static sampling at a number of locations within the room. The location of the static sampling should be in the area where the technician is liable to be working and predominantly in areas where you consider high concentrations of particulates to be likely. The areas of high concentrations will be determined by the air flows in the room which will be produced by the convection currents from the furnace, plus doors, windows and any extraction system. The vertical location should reflect, if possible, the height of the breathing zone of the technician in his most usual stance, whether seated or standing.

Sampling should be over as long a period as possible to reflect the exposure over an 8 h working day. Since large variations in exposure are possible, monitoring should be repeated for several days.

Response 7.2

In order to answer this question, you will need to combine what you have learnt from Sections 6.2.1, 6.3.1 and 7.25. Starting from the sampling position, the following will be required:

(i) A particulate filter. This may be inside the flue, or outside the flue connected by a heated pipe. Temperature measurement would be necessary to ensure that the heating is at the same temperature as the sampling temperature.

(ii) (a) An absorption train, perhaps cooled in ice–water to prevent evaporation, and a trap to protect the subsequent pump. Alternatively, you could have:

(b) A method for rapid cooling of the gases at the same time preventing condensation of water (dilution, drier or chiller), followed (after a pump and flow meter) by an instrument measuring gases at ambient temperatures. A further possibility could be:

(c) An instrument capable of taking measurements at high temperatures, followed by a cooling/drying system to protect the subsequent pump.

(iii) A pump.

(iv) A gas meter or flow meter.

There would also need to be a pitot tube to measure the gas flow at the sampling point within the duct to ensure isokinetic sampling.

Response 7.3

From a practical point of view, routine analysis will often be performed in small laboratories close to the workplace being monitored and with limited facilities. Under such circumstances, ultraviolet/visible spectrometry may be a more appropriate method. As well as a greater capital investment for an atomic absorption spectrometer, adequate ventilation is necessary, and also a regular gas cylinder supply. There would probably be an insufficient throughput of samples to justify the additional cost of ICP-OES or ICP-MS.

From an analytical point of view, particulate samples from one workplace will be of relatively constant (and known) composition. Potential interferences, which limit the use of ultraviolet/visible spectrometry for samples of unknown composition, can be readily assessed.

Response 7.4

The small sample masses of atmospheric particulates (mg or below) may mean that you are working close to the limits of detection of the available techniques. The limits of detection of each technique are different for each element (see Table 7.2) and so the most appropriate technique may differ for each analysis. In other areas of environmental analysis, the sample size may not be such a restriction and preconcentration may be used to decrease the lower limit of detection.

Response 7.5

With such a general question, I cannot put the criteria in any rank order, but they should include the following:

1. *Ease of solubility of the analyte.* If the analyte is soluble in water or dilute acid, solution analytical techniques are usually the most convenient to use.
2. *Number of elements being analysed.* You should re-read the description of the techniques to determine which are most suitable for multiple-element analysis.
3. *Availability of equipment.* Many of the solid-state techniques will only be found in laboratories dedicated to solid-state analysis.
4. *Sensitivity.* Often, you will be working close to the limits of detection of the methods. The most sensitive technique will differ for each element.
5. *Compliance with specified method.* Some legislation requires the use of specific procedures for the analysis. Other legislation accepts that alternative techniques may be used if they have suitable accuracy and reliability for the application. The validation of an alternative method may, however, be a long and costly process.

Chapter 8

Response 8.1

1. Each of the components will have different physical and chemical properties, which in turn leads to different bioconcentration ability, rates of degradation and toxicity. If a value for the total toxicity of a sample is required, this would involve determining individual concentrations and compensating for their different toxicities, for example, by using *toxic equivalent factors*.
2. The majority of analytical schemes for ng kg^{-1} concentrations of organic compounds would involve chromatographic separation of the interfering compounds (after extensive pretreatment), which in turn provides at least a partial separation of the PCDDs and PCDFs.
3. As you might expect, the relative quantities of each of the compounds will be different from each production source. Under favourable circumstances, estimation of the relative concentrations can give an indication of their likely origin.

Response 8.2

The assumption in the use of internal standards is that the standard will behave identically in the extraction to the compound being analysed. An isotopically labelled compound would be closer in behaviour than a chemically distinct compound.

A second benefit is that the labelled compound serves for peak identification – an important consideration when you remember the large number of peaks which may be found even in a selected ion chromatogram.

Response 8.3

1. The peak should occur at the expected retention time for the chromatographic column. It is easy to forget that the mass spectrometer is simply a highly sophisticated detector for the chromatograph and that retention times are a good primary means of identification.
2. The peak should be monitored at two or more m/z values, corresponding to the same molecular fragment with different distributions of ^{35}Cl and ^{37}Cl in the molecule. The relative intensities should correspond to the expected statistical distribution. A complete mass spectrum could be used to attribute the peak to a dioxin or diobenzofuran rather than an impurity. However, the fragmentation patterns of the dioxins and furans are often too similar to allow positive identification of individual members of the two groups.
3. The peak should only be considered genuine if it is at least 2.5 times greater than the background noise. Below this intensity there is a possibility that the 'peak' may simply be part of the background.

Response 8.4

You should include the following in your programme:

- Blank determinations of all batches of reagents used
- Analysis of Standards from National Laboratories
- Replicate analyses, which may be unmarked ('blind') replicates
- Repetition of one unknown sample throughput the sequence
- Frequent checks on:
  - the purity of reagents throughout the programme
  - the recovery of standards in the pretreatment
  - the resolution of the GC column

These features are little different from those which you would include in any analytical scheme (see Section 2.9), but for PCDD and PCDF analysis there are severe limitations on how many standards and replicates may be included due to the laboratory time required for each sample. Remember that it takes approximately one hour for each GC analysis, as well as the time taken in the pretreatment stage.

Response 8.5

The first of these is common to all analyses where there is extraction from a solid and has been discussed earlier in Section 5.2, i.e. there is always the uncertainty that the extraction is complete. The extraction efficiency of the compound within the sample matrix may also differ from that of the internal standard.

The second arises from the impracticability of using internal standards for all 210 PCDDs and PCDFs, and the uncertainty involved in the determination of average response factors. The practical limit is often seen as one standard per isomer group, which is usually the compound including the 2,3,7,8-substitution pattern.

Bibliography†

Introductory Material – Environmental Chemistry, Pollution and Analysis

Harris, D. C., *Quantitative Chemical Analysis*, 5th Edn, W. H. Freeman and Co., New York, 1998: ISBN 0-7167-2881-8.

Manahan, S. E., *Environmental Chemistry*, 7th Edn, Lewis Publishers, Boca Raton, FL, 1999: ISBN 1-56670-492-8.

Meyers, R. A. and Dittrich, D. H. (Eds.), *Encyclopedia of Environmental Pollution and Cleanup*, (Two-volume set), Wiley-Interscience, New York, 1999: ISBN 0-471-31612-1.

O'Neil, P., *Environmental Chemistry*, 3rd Edn, Blackie, London, 1998: ISBN 0-7514-0483-7.

Skoog, D. A., West, D. M., Holler, J. F. and Crouch, S. R., *Analytical Chemistry: An Introduction*, 7th Edn, Saunders College Publishing, Forth Worth, TX, 2000: ISBN 0-03-020293-0.

Compilations of Standard Analytical Methods

These have been produced by environmental authorities worldwide and are available in printed copy form. Methods from some authorities may also be available on websites.

Environmental Protection Agency Methods, National Technical Information Service (NTIS), Springfield, VA. These include the following:

- 500 Series – Drinking Water

† The opinions expressed within this bibliography are not those of the publisher.

- 600 Series – Waste Water
- 800 Series – Solid Waste

EPA home page – http://www.epa.org
NTIS home page – http://www.ntis.gov

Summary volume – Keith, L. H. (Ed.), *Compilation of EPA's Sampling and Analysis Methods*, 2nd Edn, CRC Press, Boca Raton, FL, 1996: ISBN 1-56670-170-8.

Manual of Analytical Methods, National Institute of Occupational Safety and Health (NIOSH), NTIS, Springfield, VA.

Methods for the Determination of Hazardous Substances, Health and Safety Executive, HMSO, London.

Methods for the Examination of Waters and Associated Materials, Standing Committee of Analysts, HMSO, London.

Standard Methods for the Examination of Water and Wastewater – published jointly by the American Public Health Association, the American Waterworks Association and the Water Environment Federation, Washington, DC.

Books, Papers and Websites†

These have been selected to:

- Give detailed guidance in specific subject areas.
- Give further background on new or unfamiliar techniques.
- Exemplify how the techniques are used in practice or how the analytical data may be used.
- Compare and contrast different techniques – the essence of this book!

Chapter 2

Barcelo, D. (Ed.), *Environmental Analysis: Techniques, Applications and Quality Assurance*, Elsevier, Amsterdam, 1993: ISBN 0-444-89648-1.

Hemond, H. F. and Fechner-Levy, E. J., *Chemical Fate and Transport in the Environment*, 2nd Edn, Academic Press, New York, 2000: ISBN 0-12-340275-1.

Howard, A. G. and Stathan, P. J., *Inorganic Trace Analysis: Philosophy and Practice*, Wiley, Chichester, UK, 1993: ISBN 0-471-94144-1.

Prichard, E. (Co-ordinating Author), *Quality in the Analytical Chemistry Laboratory*, ACOL Series, Wiley, Chichester, UK, 1995: ISBN 0-471-95470-5.

† As of August 2001. The material displayed is not endorsed by the author or the publisher.

Quevauviller, Ph. (Ed.), *Quality Assurance in Environmental Monitoring, Sampling and Sample Pretreatment*, VCH, Weinheim, 1995: ISBN 3-527-28724-8.

Chapter 3

Bartram, J. and Balance, R. (Eds), *Water Quality Monitoring*, E and F Spon, London, 1996: ISBN 0-419-22320-7. This book includes both chemical and biological monitoring techniques.

Fresenius, W., Quentin, K. E. and Scheider, W. (Eds), *Water Analysis*, Springer-Verlag, Berlin, 1988: ISBN 3-540-17723-X.

HMSO, 'General principles for sampling waters and waste materials, estimation of flow and load 1996', HMSO, London, 1996: ISBN 0-11-752364-X.

Krajca, J. M. (Ed.), *Water Sampling*, Ellis Horwood, Chichester, UK, 1989: ISBN 0-85312-813-8.

http://www.environment-agency.gov.uk

This website contains water quality information of all rivers in England and Wales based on the UK General Quality Assessment (see SAQ 3.4). Click on 'What's in your backyard'. In order to start, you will need to input a UK postcode. This has the form of one (or two) letters, one (or two) digits, space, digit and two letters. You could start at the University of Sunderland (SR1 3SD), and then navigate westwards along the River Wear to its source in the Pennine Hills. By clicking on the sampling point on the map, you can obtain historical data covering the past 10 years.

Chapter 4

Baugh, P. J., *Gas Chromatography: A Practical Approach*, IRL Press, OUP, Oxford, 1993: ISBN 0-19-963272-3. Chapter 9, of this text, 'Environmental Analysis using Gas Chromatography', includes several detailed protocols.

Bloemen, H. J. Th. and Burn, J. (Eds), *Chemistry and Analysis of Volatile Organic Compounds in the Environment*, Blackie, Glasgow, UK, 1993: ISBN 0-7514-0000-9.

Boehm, P. D., Douglas, G. S., Burns, W. A., Mankiewicz, P. J., Page, D. S. and Bence, A. E., 'Application of petroleum hydrocarbon chemical fingerprinting and allocation techniques after the Exxon Valdez oil spill', *Marine Pollution Bulletin*, **34**(8), 599–613 (1997).

Bruner, F., *Gas Chromatographic Environmental Analysis*, VCH, New York, 1993: ISBN 1-56081-011-4.

Minoia, C. and Caroli, S. (Eds), *Application of Zeeman Graphite Furnace Atomic Absorption Spectrometry in the Chemical Laboratory and in Toxicology*, Pergamon Press, Oxford, UK, 1992: ISBN 0-09-041019-7. This book includes several chapters on environmental analysis.

Smedes, F., de Jong, A. S. and Davies, I. M., 'Determination of (mono-, di- and) tri-butyltin in sediments. Analytical methods', *Journal of Environmental*

Monitoring, **2**(6), 541–549 (1999). This is a review paper of possible methodologies from sampling and storage, pretreatment, clean-up and concentration, analysis and quality assurance.

Thurman, E. M. and Mills, M. S., *Solid Phase Extraction: Principles and Practice*, Wiley-Interscience, New York, 1998: ISBN 0-471-61422-X. Chapter 7 of this text specifically concerns environmental analysis, although there are many other environmental examples given elsewhere in the book.

Chapter 5

Allen, S. E. (Ed.), *Chemical Analysis of Ecological Materials*, 2nd Edn, Blackwell Scientific Publications, Oxford, UK, 1989: ISBN 0-632-01742-2.

British Standards Institute, 'Investigation of potentially contaminated sites. Code of practice', BS 10175: 2001, BSI, London, 2001: ISBN 0-580-33090-7.

Carro, A. M., Lorenzo, R. A., Vazquez, M. J., Abuin, M. and Cela, R., 'Different extraction techniques in the preparation of methylmercury biological samples: classic extraction, supercritical fluid and microwave extraction', *International Laboratory*, 23–27 (November 1998).

Dean, J. R., *Extraction Methods for Environmental Analysis*, Wiley, Chichester, UK, 1998: ISBN 0-471-98287-3. This book describes, compares and contrasts the Soxhlet, automated Soxhlet, sonication, SFE, microwave-assisted extraction and accelerated solvent extraction techniques.

Guerin, T. F., 'The extraction of aged polycyclic aromatic hydrocarbon (PAH) residues from a clay soil using sonication and a Soxhlet procedure: a comparative study', *Journal of Environmental Monitoring*, **1**(1), 63–67 (1999).

Lopez-Avila, V., Young, R. and Teplitsky, N., 'Microwave-assisted extractions: an alternative to Soxhlet, sonication, and supercritical fluid extraction', *Journal of AOAC International*, **79**(1), 142–156 (1996).

Que Hee, S. S., *Hazardous Waste Analysis*, ABS Group Inc., Rockville, MA, 1999: ISBN 0-86587-609-6. The book is a comprehensive guide to all aspects of waste analysis.

Reid, B. J., Jones, K. C. and Semple, K. T., 'Bioavailablity of persistent organic pollutants in soils and sediments – a perspective on mechanisms, consequences and assessment', *Environmental Pollution*, **108**, 103–112 (2000). This is a discussion paper concerning whether bioavailability of a pollutant can be assessed by chemical extraction methods, as is commonly assumed.

Rifai, H. S., Bedient, P. B.and Shorr, G. L., 'Monitoring hazardous waste sites: characterisation and remediation considerations', *Journal of Environmental Monitoring*, **2**(3), 199–212 (2000). This is a review paper which considers possible methods and also includes a specific case study.

Rumford, S., Yersin, J., Hetheridge, M. and Cumming, R., 'Comparison of rapid methods of groundwater sampling using direct push probes', *Land Contamination and Reclamation*, **7**(1), 41–48 (1999).

Scottish Enterprise Environmental Development, 'How to investigate Contaminated Land: Requirements for contaminated land site investigations', Scottish Enterprise, Glasgow, UK, 1994: ISBN 0-905574-13-3.

Chapter 6

Clarke, A. G. (Ed.), *Industrial Air Pollution Monitoring*, Chapman & Hall, London, 1998: ISBN 0-412-63390-6.

Couling, S. (Ed.), *Measurement of Airborne Pollutants*, Butterworth-Heinemann, Oxford, UK, 1993: ISBN 0-7506-0885-4. This book describes the practicalities of atmospheric monitoring, using a number of monitoring programmes as case studies.

Krupa, S. V. and Legge, A. H., 'Passive sampling of ambient, gaseous air pollutants: an assessment from an ecological perspective', *Environmental Pollution*, **107**, 31–35 (2000). This paper includes comparisons of passive samplers and continuous monitors.

Lodge, Jr, J. P. (Ed.), *Methods of Air Sampling and Analysis*, 3rd Edn, Lewis, Publishers, Chelsea, MI, 1989: ISBN 0-87371-141-6.

Sigrist, M. W. (Ed.), *Air Monitoring by Spectroscopic Techniques*, Wiley-Interscience, New York, 1994: ISBN 0-471-55875-3. A detailed but readable book on remote sensing techniques.

http://www.seiph.umds.ac.uk/detr/ss_reports/ar98.htm
This site describes roadside monitoring at Marylebone Road in London and compares the various techniques used. These include the following:

- NO_2 diffusion tube and continuous NO_x analyser
- Benzene diffusion tubes versus automatic gas chromatograph
- TEOM PM_{10} versus gravimetric PM_{10}
- SO_2 bubbler versus continuous SO_2 analyser

http://www.unep.or.jp/CTT_DATA/AMON/Contents_4.html
This site describes techniques used in Japan for monitoring ambient air and stationary sources.

Chapter 7

Heal, M. R., Beverland, I. J., McCabe, M., Hepburn, W. and Angus, R. M., Intercomparison of five PM_{10} monitoring devices and the implications for exposure measurement in epidemiological research, *Journal of Environmental Monitoring*, **2**(5), 455-461(2000) (2000).

HMIP, 'Monitoring emissions of pollutants at source', Technical Guidance Note (monitoring) M2, HMSO, London, 1993: ISBN 0-117-52922-2.

Smith, S., Stribley, T., Barratt, B. and Perryman, C., Determination of PM_{10} by Partisol, TEOM, ACCU and cascade impactor instruments in the London Borough of Greenwich, *Clean Air*, **27**(3), 70–73 (1997).

http://ccar.ust.hk/% 7Ealau/epd_cdrom96/aqn
This site shows examples of monitoring equipment on site and the mobile air monitoring station in Hong Kong.

Chapter 8

HMIP, 'Determination of polychlorinated biphenyls, polychlorinated dibenzo-*p*-dioxins and polychlorinated dibenzofurans in UK soils', HMSO, London, 1989: ISBN 0-11-752268-6.

March, R. E., Splendore, M., Reiner, E. J., Mercer, R. S., Plomley, J. B., Wadell, D. S. and MacPherson, K. A., 'A comparison of three mass spectrometric techniques for the determination of dioxins/furans', *International Journal of Mass Spectrometry*, **194**, 235–246 (2000).

Sheridan, R. S. and Meola, J. R., 'Analysis of pesticide residues in fruits, vegetables and milk by gas chromatography/tandem mass spectrometry', *Journal of AOAC International*, **82**(4) 982–990 (1999).

Glossary of Terms

This section contains a glossary of terms, all of which are used in the text. It is not intended to be exhaustive, but to explain briefly those terms which often cause difficulties or may be confusing to the inexperienced reader.

Absorbance A measurement of the absorption of light; absorbance = log I_0/I, where I_0 is the intensity of the incident radiation and I is the intensity of the transmitted radiation. The operational scale is 0–1.2, but the normal working range is towards the lower half of the scale.

Absorption The process of incorporation of a gas or liquid into the bulk of a body, *or* the attenuation of light passing through a liquid.

Accuracy Closeness of a result to the true value.

Acid rain Rain with a pH of less than 5.6 (unpolluted value). The chemistry of its formation is complex but often originates from combustion processes producing SO_2/SO_3 (SO_x) and NO/NO_2 (NO_x).

Activity Concentration of ions in water after correction for their thermodynamic non-ideal behaviour. At low concentration, activity $\sim$ concentration, while at infinite dilution, activity = concentration.

Adsorption The process where molecules of a gas or liquid adhere to the surface of a body.

Aerobic Presence of oxygen, normally referring to microbial activity.

Anaerobic Absence of oxygen, normally referring to microbial activity.

Analyte The specific compound or ion within a sample which is being analysed.

Anion A negatively charged ion.

Anthropogenic Man-made.

Atmosphere The gases surrounding a body. The most obvious example is the atmosphere surrounding the earth but the term can be applied elsewhere, e.g. internal atmospheres inside buildings.

Bioconcentration Increase in concentration of a pollutant in an organism compared with the surrounding environment.

Biomagnification Increase in concentration of a pollutant in organisms along a food chain.

Buffer solution Solutions which resist a change in pH.

Carcinogen A substance (or specific forms of radiation) capable of forming cancer.

Cation A positively charged ion.

Chelating The ability of a molecule or ion to bond to a metal ion through more than one atom to form a complex. The complexes formed are more stable than similar species where the bond is through one atom, i.e. non-chelating species.

Climate change One of the observable effects which is predicted to occur from global warming. In addition to changes in temperature, there may also be changes to weather patterns, including more extreme weather conditions.

Colloid Solid particles, with sizes so small to be invisible in a simple microscope, which can be permanently suspended in water. Their suspension is stabilized by ionic forces.

Complex The chemical species formed when an ion or compound containing an atom with a lone pair of electrons (e.g. N, O and S) forms a bond to a metal ion.

Complexing agent The chemical species (an ion or a compound) which will bond to a metal ion using lone pairs of electrons.

Congener A term used with organochlorine compounds such as dioxins, PCBs, PCDFs, etc. This is a member of a group of compounds with similar structural features such as the same number of chlorines or a common substitution pattern, regardless of the total number of chlorines.

Co-precipitation The inclusion of otherwise soluble ions during the precipitation of lower-solubility species.

Degradation The breakdown of organic molecules into simpler species through a number of distinct stages. This may be by chemical or biological means.

Denitrification The process of nitrate being reduced to nitrogen by microorganisms in the absence of oxygen.

Diffuse source A source of discharge of a pollutant which occurs over a wide area e.g. methane emissions from a waste site.

Ecosystem Plant and animals in an area of the environment together with that part of the physical environment relevant to their well-being.

Eluent The mobile liquid phase in liquid or ion chromatography or in solid-phase extraction.

Endocrine disruptor A compound which can interfere with the maintenance of normal blood hormone levels or the subsequent action of these hormones.

Eutrophication The enrichment of a water body with nutrients which lead to excessive plant growth and ultimately death of the ecosystem.

Filter feeders Organisms which feed by ingesting small food particles from the surrounding water, e.g. shellfish.

Flue gas Gas emitted from a chimney – also known as stack gas.

Fossil fuel Hydrocarbon fuel derived from fossil remains. It includes coal, oil and natural gas.

Fulvic acid Naturally occurring high-molecular-mass organic compounds which are soluble at all pH levels.

Global warming The predicted increase in the average world temperature due to increase of greenhouse gases in the atmosphere, coupled with destruction of the world's forests. Such predictions are based on average temperatures. Some areas may cool, while others increase in temperature.

Greenhouse effect The effect caused by certain molecules (notably carbon dioxide and water) in the atmosphere to absorb and re-radiate infrared radiation. In an unpolluted atmosphere, this is a natural effect which maintains the average atmospheric temperature. With an increase in concentration of greenhouse gases, this average temperature is predicted to rise.

Greenhouse gas A compound which will absorb infrared radiation and so contribute to any greenhouse effect. The molecules of the compound have to contain at least two atoms and, if diatomic, more than one element.

Groundwater Sub-surface water in soils and geological formations where the ground has become saturated with water.

Humic acid Naturally occurring high-molecular-mass organic compounds which are acid-soluble but are precipitated by base.

Hydrophilic The term applied to compounds, molecules, or sometimes parts of molecules, which have an affinity for water. This affinity may be due to the presence of polar atoms such as oxygen or nitrogen, or may be due to ionic groups.

Hydrophobic The term applied to compounds, molecules or sometimes parts of molecules, which lack any affinity for water.

Hydrosphere The different forms of water found on or in the earth – oceans, seas, lakes, rivers, groundwater, glaciers, etc.

Ion exchange The exchange of ions of the same charge between water and a solid in contact with it.

Ion trap A form of separation and detection in mass spectrometry in which the charged ions are stored in closed orbits and are selectively extracted for detection and quantification.

Isocratic The term used for an eluent in liquid chromatography which remains of constant composition. This is in contrast to some advanced forms of chromatography which optimize separations by a continuous change in eluent composition, known as gradient (rather than isocratic) elution.

Kuderna–Danish evaporator Apparatus for sample concentration consisting of a small (10 ml) graduated test tube connected directly beneath a 250 or

500 ml flask. A steam bath provides heat for evaporation with the concentrate collecting in the test tube.

Leachate The liquid after passing through a substance which contains soluble extracts.

Least squares A method of determining the best straight line to fit a series of points on a graph. This 'best line' minimizes the square of the distances of the points to the line as measured along the y-axis – the direction in which the greatest experimental error is expected. The calculation is a standard feature of scientific calculators and spreadsheets.

Lipid High-molecular-mass organic compound which can be extracted into organic solvents.

Lithosphere The crust and mantle of the earth's surface. This is the section of the earth which is of greatest relevance when discussing environmental effects.

Lysimeter A device for collecting water from the pore spaces of soils and for determining the soluble constituents removed by drainage.

Metabolism Chemical reactions which take place within a living organism.

Nitrification Oxidation of ammonia to nitrite and then to nitrate by the action of microorganisms.

Organometallic An organic compound in which a metal is covalently bonded to carbon.

Oxidation Addition of oxygen to a molecule, the removal of electrons or the removal of hydrogen from a molecule.

Ozone depletion The destruction of ozone molecules by radicals formed by molecules which are stable enough in the lower atmosphere to be transportable to the upper atmosphere. The molecules often contain a halogen atom. Nitrous oxide, N_2O, can also cause ozone depletion.

Ozone layer A section of the upper atmosphere between 12–60 km which contains a higher level of ozone than other sections. This section protects the earth from UV radiation between 220 and 330 nm.

Pesticide A synthetic compound used to control the number or spread of organisms. The main types are insecticides (insect control), herbicides (weed control) and fungicides (control of fungal growth on crops, in industrial process plant or in commercial products).

Photochemical smog An oxidizing haze produced from vehicle exhausts in large conurbations during daylight hours and under thermal-inversion conditions.

Photolysis Breakdown of a compound by the action of light.

Photosynthesis The chemical process by which green plants synthesize organic compounds from carbon dioxide and water using sunlight as an energy source.

Point source A discharge which can be readily identified and located.

Precipitation Formation of an insoluble salt by mixing of two previously soluble ions, *or* rainfall.

Precision Reproducibility of an analytical result.

Protocol Formal statement of an analytical procedure.

Radical An atom or molecule containing an unpaired electron.

Reduction The removal of oxygen from a molecule, the addition of electrons or the addition of hydrogen to a molecule.

Remote sensing Gathering and recording information by techniques which do not involve direct contact.

Saponification Breakdown of fat using alkali. The products are carboxylate salts and alcohols.

Sediment Small particles of mineral or organic matter on river or sea beds.

Stack gas Gas emitted from a chimney – also known as flue gas.

Stratosphere The portion of the atmosphere between 10 and 50 km altitude.

Thermal inversion Atmospheric conditions which produce a layer of cold gas below a layer of warmer gas (the opposite condition to what is normally found). This produces very placid and stable atmospheric conditions which allow pollutants to build up and inter-react.

Vadose zone Soil containing water but not to saturation level.

Units of Measurement and Physical Constants

There is a bewildering array of measurement units in common use. The units used to describe water, atmospheres and solids have developed independently and in each of these areas there may be more than one system in frequent use. SI units, which have been recommended by many international scientific bodies, are unfortunately not often used in Environmental Science. This section begins with a description of some of the units likely to be found in the environmental literature and the conventions used in this present textbook. Details of SI units are then presented.

Practical Units Used in Environmental Analysis

The approach adopted in this book is to use the units most commonly employed in environmental literature, but wherever possible to chose units based on the following:

- Mass of analyte/unit volume — Water, Atmospheres
- Mass of analyte/unit mass — Solids

Typical units would then be:

- Water — mg l^{-1}, μg l^{-1}
- Atmosphere — mg m^{-3}, μg m^{-3}
- Solids — mg kg^{-1}, μg kg^{-1}

An alternative system of units is sometimes found in the environmental literature which is based on parts per million (ppm), parts per billion (ppb), and sometimes parts per trillion (ppt). These are avoided wherever possible owing to possible ambiguities in their interpretation. As shown below, different definitions are used for the terms in liquids and solids, and in gas analysis.

For liquids and solids:

$$\begin{aligned} \text{ppm} &= \text{parts per million (mass/mass)} \\ &= \text{mg kg}^{-1} \\ &\simeq \text{mg l}^{-1} \text{(assuming density of sample} \simeq \text{1g ml}^{-1}) \end{aligned}$$

Similarly:

$$\begin{aligned} \text{ppb} &\simeq \mu\text{g l}^{-1} \\ \text{ppt} &\simeq \text{ng l}^{-1} \end{aligned}$$

For gases:

$$\begin{aligned} \text{ppm} &= \text{parts per million (volume/volume)} \\ &= \mu\text{l l}^{-1} \\ \text{ppb} &= \text{nl l}^{-1} \\ \text{ppt} &= \text{pl l}^{-1} \end{aligned}$$

You should also note that billion and trillion always follow the US rather than the UK usage, i.e.

$$\begin{aligned} 1 \text{ billion} &= 10^9 \\ 1 \text{ trillion} &= 10^{12} \end{aligned}$$

SI Units

Base SI units and physical quantities

Quantity	Symbol	SI unit	Symbol
length	l	metre	m
mass	m	kilogram	kg
time	t	second	s
electric current	I	ampere	A
thermodynamic temperature	T	kelvin	K
amount of substance	n	mole	mol
luminous intensity	I_v	candela	cd

Prefixes used for SI units

Factor	Prefix	Symbol
10^{21}	zetta	Z
10^{18}	exa	E
10^{15}	peta	P
10^{12}	tera	T
10^{9}	giga	G
10^{6}	mega	M
10^{3}	kilo	k
10^{2}	hecto	h
10	deca	da
10^{-1}	deci	d
10^{-2}	centi	c
10^{-3}	milli	m
10^{-6}	micro	μ
10^{-9}	nano	n
10^{-12}	pico	p
10^{-15}	femto	f
10^{-18}	atto	a
10^{-21}	zepto	z

Derived SI units with special names and symbols

Physical quantity	SI unit		Expression in terms of base or derived SI units
	Name	Symbol	
frequency	hertz	Hz	$1 \text{ Hz} = 1 \text{s}^{-1}$
force	newton	N	$1 \text{ N} = 1 \text{ kg m s}^{-2}$
pressure; stress	pascal	Pa	$1 \text{ Pa} = 1 \text{ N m}^{-2}$
energy; work; quantity of heat	joule	J	$1 \text{ J} = 1 \text{ N m}$
power	watt	W	$1 \text{ W} = 1 \text{ J s}^{-1}$
electric charge; quantity of electricity	coulomb	C	$1 \text{ C} = 1 \text{ A s}$
electric potential; potential difference; electromotive force; tension	volt	V	$1 \text{ V} = 1 \text{ J C}^{-1}$
electric capacitance	farad	F	$1 \text{ F} = 1 \text{ C V}^{-1}$
electric resistance	ohm	Ω	$1 \Omega = 1 \text{ V A}^{-1}$

(continued overleaf)

Derived SI units with special names and symbols *(continued)*

Physical quantity	SI unit		Expression in terms of base
	Name	Symbol	or derived SI units
electric conductance	siemens	S	$1\ S = 1\Omega^{-1}$
magnetic flux; flux of magnetic induction	weber	Wb	$1\ Wb = 1\ V\ s$
magnetic flux density; magnetic induction	tesla	T	$1\ T = 1\ Wb\ m^{-2}$
inductance	henry	H	$1\ H = 1\ Wb\ A^{-1}$
Celsius temperature	degree Celsius	°C	$1°C = 1\ K$
luminous flux	lumen	lm	$1\ lm = 1\ cd\ sr$
illuminance	lux	lx	$1\ lx = 1\ lm\ m^{-2}$
activity (of a radionuclide)	becquerel	Bq	$1\ Bq = 1\ s^{-1}$
absorbed dose; specific energy	gray	Gy	$1\ Gy = 1\ J\ kg^{-1}$
dose equivalent	sievert	Sv	$1\ Sv = 1\ J\ kg^{-1}$
plane angle	radian	rad	1[a]
solid angle	steradian	sr	1[a]

[a] rad and sr may be included or omitted in expressions for the derived units.

Physical Constants

Recommended values of selected physical constants[a]

Constant	Symbol	Value
acceleration of free fall (acceleration due to gravity)	g_n	$9.806\ 65\ m\ s^{-2}$ [b]
atomic mass constant (unified atomic mass unit)	m_u	$1.660\ 540\ 2(10) \times 10^{-27}\ kg$
Avogadro constant	L, N_A	$6.022\ 136\ 7(36) \times 10^{23}\ mol^{-1}$
Boltzmann constant	k_B	$1.380\ 658(12) \times 10^{-23}\ J\ K^{-1}$
electron specific charge (charge-to-mass ratio)	$-e/m_e$	$-1.758\ 819 \times 10^{11}\ C\ kg^{-1}$
electron charge (elementary charge)	e	$1.602\ 177\ 33(49) \times 10^{-19}\ C$
Faraday constant	F	$9.648\ 530\ 9(29) \times 10^{4}\ C\ mol^{-1}$

Recommended values of selected physical constants[a] *(continued)*

Constant	Symbol	Value
ice-point temperature	T_{ice}	273.15 K [b]
molar gas constant	R	8.314 510(70) J K^{-1} mol^{-1}
molar volume of ideal gas (at 273.15 K and 101 325 Pa)	V_m	22.414 10(19) $\times 10^{-3}$ m^3 mol^{-1}
Planck constant	h	6.626 075 5(40) $\times 10^{-34}$ J s
standard atmosphere	atm	101 325 Pa[b]
speed of light in vacuum	c	2.997 924 58 $\times 10^8$ m s^{-1} [b]

[a] Data are presented in their full precision, although often no more than the first four or five significant digits are used; figures in parentheses represent the standard deviation uncertainty in the least significant digits.

[b] Exactly defined values.

The Periodic Table

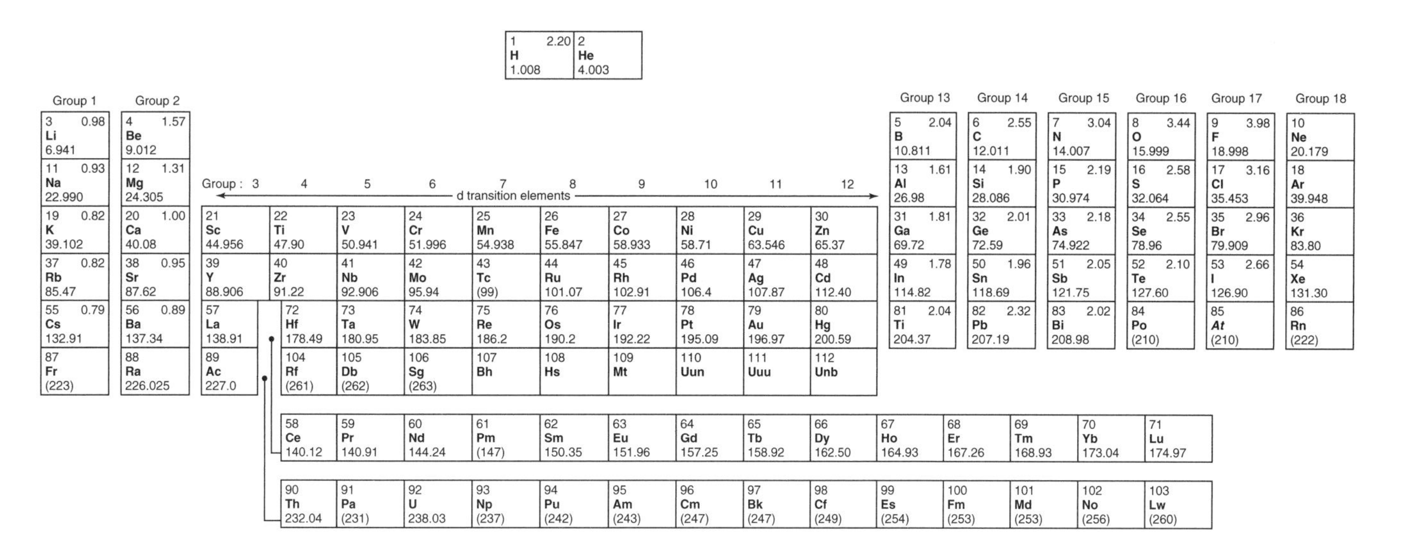

0.98	Pauling electronegativity
3	Atomic number
Li	Element
6.941	Atomic weight (^{12}C)

1 2.20 **H** 1.008	2 **He** 4.003

d transition elements: Group 3 to 12

Group 1	Group 2	Group: 3	4	5	6	7	8	9	10	11	12	Group 13	Group 14	Group 15	Group 16	Group 17	Group 18
3 0.98 **Li** 6.941	4 1.57 **Be** 9.012											5 2.04 **B** 10.811	6 2.55 **C** 12.011	7 3.04 **N** 14.007	8 3.44 **O** 15.999	9 3.98 **F** 18.998	10 **Ne** 20.179
11 0.93 **Na** 22.990	12 1.31 **Mg** 24.305											13 1.61 **Al** 26.98	14 1.90 **Si** 28.086	15 2.19 **P** 30.974	16 2.58 **S** 32.064	17 3.16 **Cl** 35.453	18 **Ar** 39.948
19 0.82 **K** 39.102	20 1.00 **Ca** 40.08	21 **Sc** 44.956	22 **Ti** 47.90	23 **V** 50.941	24 **Cr** 51.996	25 **Mn** 54.938	26 **Fe** 55.847	27 **Co** 58.933	28 **Ni** 58.71	29 **Cu** 63.546	30 **Zn** 65.37	31 1.81 **Ga** 69.72	32 2.01 **Ge** 72.59	33 2.18 **As** 74.922	34 2.55 **Se** 78.96	35 2.96 **Br** 79.909	36 **Kr** 83.80
37 0.82 **Rb** 85.47	38 0.95 **Sr** 87.62	39 **Y** 88.906	40 **Zr** 91.22	41 **Nb** 92.906	42 **Mo** 95.94	43 **Tc** (99)	44 **Ru** 101.07	45 **Rh** 102.91	46 **Pd** 106.4	47 **Ag** 107.87	48 **Cd** 112.40	49 1.78 **In** 114.82	50 1.96 **Sn** 118.69	51 2.05 **Sb** 121.75	52 2.10 **Te** 127.60	53 2.66 **I** 126.90	54 **Xe** 131.30
55 0.79 **Cs** 132.91	56 0.89 **Ba** 137.34	57 **La** 138.91	72 **Hf** 178.49	73 **Ta** 180.95	74 **W** 183.85	75 **Re** 186.2	76 **Os** 190.2	77 **Ir** 192.22	78 **Pt** 195.09	79 **Au** 196.97	80 **Hg** 200.59	81 2.04 **Ti** 204.37	82 2.32 **Pb** 207.19	83 2.02 **Bi** 208.98	84 **Po** (210)	85 ***At*** (210)	86 **Rn** (222)
87 **Fr** (223)	88 **Ra** 226.025	89 **Ac** 227.0	104 **Rf** (261)	105 **Db** (262)	106 **Sg** (263)	107 **Bh**	108 **Hs**	109 **Mt**	110 **Uun**	111 **Uuu**	112 **Unb**						

58 **Ce** 140.12	59 **Pr** 140.91	60 **Nd** 144.24	61 **Pm** (147)	62 **Sm** 150.35	63 **Eu** 151.96	64 **Gd** 157.25	65 **Tb** 158.92	66 **Dy** 162.50	67 **Ho** 164.93	68 **Er** 167.26	69 **Tm** 168.93	70 **Yb** 173.04	71 **Lu** 174.97
90 **Th** 232.04	91 **Pa** (231)	92 **U** 238.03	93 **Np** (237)	94 **Pu** (242)	95 **Am** (243)	96 **Cm** (247)	97 **Bk** (247)	98 **Cf** (249)	99 **Es** (254)	100 **Fm** (253)	101 **Md** (253)	102 **No** (256)	103 **Lw** (260)

Index

Bold indicates a reference to an analytical method or methods. Italic indicates an explanation in the *Glossary of Terms*.